Lecture Notes in Computer Science   9770

Commenced Publication in 1973
Founding and Former Series Editors:
Gerhard Goos, Juris Hartmanis, and Jan van Leeuwen

More information about this series at http://www.springer.com/series/8851

Ngoc Thanh Nguyen · Ryszard Kowalczyk
Joaquim Filipe (Eds.)

Transactions on Computational Collective Intelligence XXIV

 Springer

Editors-in-Chief

Ngoc Thanh Nguyen
Faculty of Computer Science
  and Management
Wrocław University of Technology
Wrocław
Poland

Ryszard Kowalczyk
Swinburne University of Technology
Hawthorn
Australia

Guest Editor

Joaquim Filipe
Escola Superior de Tecnologiy de Setébual
Setúbal
Portugal

ISSN 0302-9743   ISSN 1611-3349 (electronic)
Lecture Notes in Computer Science
ISBN 978-3-662-53524-0   ISBN 978-3-662-53525-7 (eBook)
DOI 10.1007/978-3-662-53525-7

Library of Congress Control Number: 2016953315

Printed on acid-free paper

This Springer imprint is published by Springer Nature
The registered company is Springer-Verlag Berlin Heidelberg
The registered company address is: Heidelberger Platz 3, 14197 Berlin, Germany

Transactions on Computational Collective Intelligence XXIV

Preface

The present special issue of *Transactions on Computational Collective Intelligence* (TCCI) includes extended and revised versions of a set of selected papers from the International Joint Conference on Computational Intelligence – IJCCI 2013 and IJCCI 2015.

The purpose of IJCCI is to bring together researchers, engineers, and practitioners interested in several areas of computational intelligence, including theory and applications of evolutionary computing, fuzzy systems, and neural networks.

After a strict reviewing process, three papers from IJCCI 2013 and six papers from IJCCI 2015 were selected for this volume of TCCI, encompassing relevant topics of current research on computational intelligence.

Particle swarms continue to attract research efforts, as exemplified by two of the selected papers: "Dynamic Topologies for Particle Swarms" authored by Carlos M. Fernandes, J.L.J. Laredo, J.J. Merelo, C. Cotta, and A.C. Rosa, and "Evaluative Study of PSO/Snake Hybrid Algorithm and Gradient Path Labeling for Calculating Solar Differential Rotation" authored by Ehsan Shahamatnia, André Mora, Ivan Dorotovič, Rita A. Ribeiro, and José M. Fonseca.

We selected three papers that presented relevant research work on evolutionary optimization and genetic programming, namely, "The Uncertainty Quandary: A Study in the Context of the Evolutionary Optimization in Games and Other Uncertain Environments" authored by Juan J. Merelo et al., "Hybrid Single Node Genetic Programming for Symbolic Regression" authored by Jiri Kubalik, Eduard Alibekov, Jan Zegklitz, and Robert Babuska, and a paper about Lindenmayer systems (L-systems) entitled "L2 Designer: A Tool for Genetic L-system Programming in Context of Generative Art," authored by Tomáš Konrády, Kamila Štekerová, and Barbora Tesařová.

The field of machine learning is another hot topic that deserves plenty of attention from the research community on computational intelligence and we selected three papers that present different applications of machine learning, including a paper on developmental robotics using humanoid robots, entitled "Manifold Learning Approach Toward Constructing State Representation for Robot Motion Generation," authored by Yuichi Kobayashi and Ryosuke Matsui, a paper describing applied research to functional magnetic resonance imaging (fMRI) entitled "The Existence of Two Variant Processes in Human Declarative Memory: Evidence Using Machine Learning Classification Techniques in Retrieval Tasks," by Alex Frid, Hananel Hazan, Ester Koilis, Larry M. Manevitz, Maayan Merhav, and Gal Star, and also a paper involving time series forecasting, entitled "Divide and Conquer Ensemble Method for Time Series Forecasting," authored by Jan Kostrzewa, Giovanni Mazzocco, and Dariusz Plewczynski.

Finally, we concluded our selection with a paper that presents a survey of a new research area, ephemeral computing, related to bioinspired optimization, evolutionary computation, complex systems, and autonomic computing. This paper, entitled "Application Areas of Ephemeral Computing: A Survey," was authored by Carlos Cotta et al. and is another good example of the application focus of this conference, without forgetting the importance of theoretical aspects because, as Ludwig Boltzmann taught us, "there is nothing more practical than a good theory."

We would like to thank all the authors for their contributions and also the reviewers for their time and expertise. Finally, we would also like to express our gratitude to the LNCS editorial staff of Springer and in particular to Prof. Ryszard Kowalczyk for his patience and availability during this process.

July 2016   Joaquim Filipe

Organization

Transactions on Computational Collective Intelligence

This Springer journal focuses on research on the applications of the computer-based methods of computational collective intelligence (CCI) and their applications in a wide range of fields such as the Semantic Web, social networks, and multi-agent systems. It aims to provide a forum for the presentation of scientific research and technological achievements accomplished by the international community.

The topics addressed by this journal include all solutions to real-life problems for which it is necessary to use CCI technologies to achieve effective results. The emphasis of the papers is on novel and original research and technological advancements. Special features on specific topics are welcome.

Contents

Dynamic Topologies for Particle Swarms

Carlos M. Fernandes[1,2(✉)], J.L.J. Laredo[2], J.J. Merelo[2], C. Cotta[3],
and A.C. Rosa[1]

[1] LARSyS: Laboratory for Robotics and Systems in Engineering and Science,
University of Lisbon, Lisbon, Portugal
{cfernandes,acrosa}@laseeb.org
[2] Department of Architecture and Computer Technology,
University of Granada, Granada, Spain
juanlu.jimenez@gmai.com, jmerelo@gmail.com
[3] Departamento de Lenguages y Ciencias de la Computación,
University of Malaga, Malaga, Spain
ccottap@lcc.uma.es

Abstract. The Particle Swarm Optimization (PSO) algorithm is a population-based metaheuristics in which the individuals communicate through decentralized networks. The network can be of many forms but traditionally its structure is predetermined and remains fixed during the search. This paper investigates an alternative approach. The particles are positioned on a 2-dimensional grid of nodes. During the run, they move through the network according to simple rules, while interacting with each other using signs that they leave on the nodes. The links between the particles – and consequently the information flow – are then defined at each time step by the position of the particle on the grid. As a result, each particle's set of neighbors and connectivity degree varies during the search progress. The particles can move randomly or instead track signs left by other particles on the grid. In this paper, after a formal description of the general model, two different strategies (random and sign-based) are tested and compared to standard topologies on unimodal and multimodal functions, including a rotated and a shifted function with noise from the CEC benchmark. The experiments demonstrate that the dynamics provided by the proposed structure results in a more consistent and stable performance throughout the test set. The working mechanisms of the model are simple and easy to implement.

1 Introduction

The Particle Swarm Optimization (PSO) is an optimization algorithm inspired by the social behavior of bird flocks and fish schools [5]. PSO search is performed by a swarm of candidate solutions (called *particles*) that move around a fitness landscape guided by mathematical rules that define their velocity and position at each time step. Each particle's velocity vector is influenced by its best known position and by the best known positions of its neighbors. The neighborhood of each particle – and conse-quently the flow of information throughout the population – is defined *a priori* by a

© Springer-Verlag Berlin Heidelberg 2016
N.T. Nguyen et al. (Eds.): TCCI XXIV, LNCS 9770, pp. 1–18, 2016.
DOI: 10.1007/978-3-662-53525-7_1

population topology. Therefore, the chosen structure deeply affects the convergence skills of the algorithm.

The particles are interconnected so that they acquire information on the regions explored by other particles. In fact, it has been claimed that the distinctiveness of the algorithm lies in the interactions of the particles [7]. These networks of individuals may be of any possible structure, from sparse to dense (or even fully connected) graphs, with different degrees of connectivity and clustering in between. The most commonly used PSO population structures are the *lbest* (which connects the individuals to a local neighborhood) and the *gbest* (in which each particle is connected to every other individual). These topologies are well-studied and the major conclusions are that *gbest* is fast but frequently trapped by local optima, while *lbest* is slower but converges more often to the neighborhood of the global optima.

Since the first experiments with these topologies, researchers have tried to design structures that hold both *lbest* and *gbest* qualities. Other studies try to understand what makes a good structure. In [7], for instance, Kennedy and Mendes investigate several types of topologies and recommend the use of a lattice with von Neumann neighborhood (resulting in a connectivity degree between that of *lbest* and *gbest*).

In the proposed topology, n particles are placed randomly in a 2-dimensional grid with $q \times s$ nodes where $q \times s \geq \mu$ and μ is the population size of the algorithm. A simple set of rules guides the transit of the particles through the nodes. Every time-step, each individual checks its von Neumann neighborhood and, as in the standard PSO, updates its velocity and position using the information given by the neighbors. However, while the connectivity degree of the von Neumann structure is $k = 5$, in the proposed topology some of the particle's neighboring nodes may be empty at that given time step. Consequently, the connectivity degree is variable in the range $1 \leq k \leq 5$. Furthermore, the structure is dynamic: in each time-step, every particle updates its position on the grid (which is a different concept from the position of the particle in the fitness landscape) according to a pre-defined rule. This rule, which is implemented locally and without any knowledge on the global state of the system, can be based on stigmergy [3] – the particles leave signals on the nodes, which are followed by other particles – or on Brownian motion [8] – the particles choose the destination node randomly from the neighboring empty sites.

In Fernandes *et al.* [1], preliminary tests with Brownian motion demonstrated the validity of the approach. In this paper, a formal description of the model is given, the set of test functions is enlarged, the signal-based configuration is tested and compared to the Brownian version and the effects of the grid size in the performance are studied. The results show that the dynamic structures clearly improve the performance of standard configurations. Furthermore, the proposed structure performs more consistently than the other topologies. Grid size affects the performance but a 1:2 ratio between the particles and nodes seems to be a good design choice for several types of optimization problems. We believe that these results, together with the simplicity of the approach and its potential as a basis for more complex movement rules validate the proposed study.

The present work is organized as follows. The next section describes PSO and its topologies, while giving a general overview on previous studies of population structures.

Section 3 describes the proposed model. Section 4 describes the experiments and discuses the results. Finally, Sect. 5 concludes the paper and outlines future lines of research.

2 Particle Swarms and Population Structure

PSO is described by a simple set of equations that define the velocity and position of each particle. The position vector of the *i-th* particle is given by $\vec{X}_i = (x_{i,1}, x_{i,2}, \dots x_{1,D})$, where D is the dimension of the search space. The velocity is given by $\vec{V}_i = (v_{i,1}, v_{i,2}, \dots v_{1,D})$. The particles are evaluated with a fitness function $f(\vec{X}_i)$ in each time step and then their positions and velocities are updated by:

$$v_{i,d}(t) = v_{i,d}(t-1) + c_1 r_1 \left(p_{i,d} - x_{i,d}(t-1)\right) + c_2 r_2 \left(p_{g,d} - x_{i,d}(t-1)\right) \qquad (1)$$

$$x_{i,d}(t) = x_{i,d}(t-1) + v_{i,d}(t) \qquad (2)$$

were p_i is the best solution found so far by particle i and p_g is the best solution found so far by the neighborhood of the particle. Parameters r_1 and r_2 are random numbers uniformly distributed in the range [0, 1] and c_1 and c_2 are acceleration coefficients that tune the relative influence of each term of the formula. The first term, influenced by the particle's best solution found so far, is known as the *cognitive part*, since it relies on the particle's own experience. The last term is the *social part*, since it describes the contribution of the community to the velocity of the particle.

In order to prevent particles from stepping out of the limits of the search space, the positions $x_{i,d}(t)$ of the particles are limited by constants that, in general, correspond to the domain of the problem: $x_{i,d}(t) \in [-Xmax, Xmax]$. Velocity may also be limited within a range in order to prevent uncontrolled growth of the velocity vector: $v_{i,d}(t) \in [-Vmax, Vmax]$. Usually, $Xmax = Vmax$.

For achieving a better balance between local and global search, Shi and Eberhart [15] introduced the inertia weight ω as a multiplying factor of the first term of Eq. 1. The modified velocity equation is:

$$v_{i,d}(t) = \omega \cdot v_{i,d}(t-1) + c_1 r_1 \left(p_{i,d} - x_{i,d}(t-1)\right) + c_2 r_2 \left(p_{g,d} - x_{i,d}(t-1)\right) \qquad (3)$$

By adjusting ω (usually within the range [0, 1.0]) together with the constants c_1 and c_2, it is possible to balance exploration and exploitation abilities of the PSO (please refer to [14] for a survey on exploration and exploitation in evolutionary algorithms). This paper uses PSOs with inertia weight.

The neighborhood of the particle (which defines in each time-step the value of p_g) is a key factor in the performance of PSO. Most of PSOs use one of two simple sociometric principles for defining the neighborhood network. One connects all the members of the swarm to one another, and it is called *gbest*, were g stands for *global*. The degree of connectivity of *gbest* is $k = n$, where n is the number of particles.

The other typical configuration, called *lbest* (where l stands for local), creates a neighborhood that comprises the particle itself and its k nearest neighbors. The most common *lbest* topology is the ring structure: particles are arranged in a ring structure (resulting in a degree of connectivity $k = 3$, including the particle).

As stated above, the topology of the population affects the performance of the PSO and it must be chosen according to the target-problem. Furthermore, each topology has its own typical behavior and its choice may also depend on the objectives or tolerance of the optimization process. Since all the particles are connected to every other and information spreads easily through the network, the *gbest* topology is known to converge fast but unreliably (it often converges to local optima).

The *lbest* converges slower than the *gbest* structure because information spreads slower through the network. However, and for the same reason, it is also less prone to converge prematurely to local optima. In-between the ring structure with $k = 3$ and the *gbest* with $k = \mu$ (where μ is the population size) there are several types of structures, each one with its advantages for a certain type of fitness landscape. However, sometimes it is not possible to assure the best configuration: the structure of the problem may be unknown, or the time requirements do not permit preliminary tests. Therefore, the research community has dedicated substantial efforts on studying the properties of population topologies for PSO.

Kennedy and Mendes [7] published an exhaustive study on topologies for PSOs. They tested several types, including *lbest*, *gbest* and von Neumann configuration with radius 1 (also kown as $L5$ neighborhood). They also tested populations arranged in randomly generated graphs. The authors conclude that when the configurations are ranked by the performance the structures with $k = 5$ (like the $L5$) perform better, but when ranked according to the number of iterations needed to meet the criteria, configurations with higher degree of connectivity perform better. These results are consistent with the premise that low connectivity favors robustness, while higher connectivity favors convergence speed (at the expense of reliability). Amongst the large set of graphs tested in [7], the Von Neumann with radius 1 configuration performed more consistently and the authors recommend its use.

Alternative topologies that combine the characteristics of standard topologies or introduce some kind of dynamics in the connections have been also proposed. In [11], Parsopoulos and Vrahatis proposed a unified PSO (UPSO) which combines both *gbest* and *lbest* configurations. Equation 1 is modified in order to include a term with p_g and a term with p_i. A parameter balances the weight of each term. The authors argue that the proposed scheme exploits the good properties of *gbest* and *lbest*. The same algorithm was later applied to dynamic optimization problems [12].

Peram *et al.* [13] proposed the fitness–distance-ratio-based PSO (FDR-PSO). The neighborhood of a particle is defined as the set of its k closest particles in the population (measured in Euclidean distance). A selective scheme is also included: the particle selects near particles that have also visited a position of higher fitness. The algorithm is compared to a standard PSO and the authors claim that FDR-PSO performs better on several test functions. However, the FDR-PSO is compared only to a *gbest* configuration, which is known to converge frequently to local optima in the majority of the functions of the test set.

A comprehensive-learning PSO (CLPSO) was proposed in [10]. Its learning strategy abandons the global best information and introduces a complex and dynamic scheme that uses all other particles' past best information. CLPSO can significantly improve the performance of the original PSO on multimodal problems.

More recently, Ni et al. [9] proposed a dynamic probabilistic PSO. The authors generate random topologies for the PSO that they use at different stages of the search. According to the authors, their strategy achieves better results than traditional static population topologies. In 2015, Augusto et al. [1] proposed a dynamic topology for PSO using probabilistic methods to choose which particles communicate and update in each time-step. The algorithm was applied to a nuclear engineering problem and the authors report that the proposed method gives better solutions than other metaheuristics.

Other strategies deal with the population in a centralized manner. For instance, in [4], the PSO varies the size of the swarm during the run, while running a solution-sharing scheme that, like in [10], uses the past best information from every particle.

The present work uses a 2-dimensional framework to force a dynamic behavior in the population structure and variability in the connectivity degree. The main objective is to search for a good compromise between high and low connectivity schemes, using dynamic connections and local interactions provided by the supporting framework. Since the Von Neumann configuration was recommended in [7], we use it as a base-structure. The following section gives a formal description of the proposed network and presents the transition rules that define the model.

3 Dynamic Structures

Let us consider a rectangular grid G of size $q \times s \geq \mu$, where μ is the size of the population of any population-based metaheuristics or model. Each node G_{uv} of the grid is a tuple $\langle \eta_{uv}, \zeta_{uv} \rangle$, where $\eta_{uv} \in \{1, \ldots, \mu\} \cup \{\bullet\}$ and $\zeta_{uv} \in (D \times N) \cup \{\bullet\}$ for some domain D. The value η_{uv} indicates the index of the individual that occupies the position $\langle u, v \rangle$ in the grid. If $\eta_{uv} = \bullet$ then the corresponding position is empty. However, that same position may still have information, namely a sign (or clue) ζ_{uv}. If $\zeta_{uv} = \bullet$ then the position is empty and unsigned. Please note that when $q \times s = \mu$, the topology is a static 2-dimensional lattice and when $q \times s = \mu$ and $q = s$ the topology is the standard square grid graph.

In the case of a PSO, the signs are placed by particles that occupied that position in the past and they consist of information about those particles, like their fitness ζ_{uv}^f or position in the fitness landscape, as well as a time stamp ζ_{uv}^t that indicates the iteration in which the mark was placed. The signs have a lifespan of L iterations, after which they are deleted. In this paper, L has been set to 1, i.e., the signs only remain in the habitat for an iteration. An investigation on the effects of longer lifespan on the dynamics of the particles is intended for a future work.

Initially, $G_{uv} = (\bullet, \bullet)$ for all $\langle u, v \rangle$. Then, the particles are placed randomly on the grid (only one particle per node). Afterwards, all particles are subject to a movement phase (or grid position update), followed by a PSO phase. The process (position update and PSO phase) repeats until a stop criterion is met.

The PSO phase is the standard iteration of a PSO, comprising position and velocity update. The only difference to a static structure is that in this case a particle may find empty nodes in its neighbourhood.

In the position update phase, each individual moves to an adjacent empty node. Adjacency is defined by the Moore neighborhood of radius r, so an individual i at $\rho_g(i) = \langle u, v \rangle$ can move to an empty node $\langle u', v' \rangle$ for which $L_\infty(\langle u, v \rangle, \langle u', v' \rangle) \leq r$. If empty positions are unavailable, the individual stays in the same node. Otherwise, it picks a neighboring empty node according to the signs on them. If there are no signs, the destination is chosen randomly amongst the free nodes. The described behavior, in which the parts of a system communicate with one another indirectly by modifying and sensing the environment, is called stigmergy, a term proposed by Grassé [3]. However, if the signals are neutral or null, the dynamics is reduced to Brownian movement. Therefore, within the framework there are two possibilities for the position update phase: stimergic, whereby the individuals place and look for signs; Brownian, whereby the individual disregards the signs and randomly selects an empty neighbor.

For the first option, let $\mathcal{N}\langle u, v \rangle = \{\langle u^{(1)}, v^{(1)} \rangle, \ldots, \langle u^w, v^w \rangle\}$. be the collection of empty neighboring nodes and let i be the individual to move. Then, the individual attempts to move to a node whose sign is as close as possible to the global optima. (Other strategies are possible but in this paper we have restricted the investigation to a *follow the best* policy.) If there are no signs in the neighborhood, the individual moves to an adjacent cell picked at random. In the alternative Brownian policy, the individual moves to an adjacent empty position picked at random. In either case, the process is repeated for the whole population. Table 1 gives the pseudo-code of a PSO with sign-based update phase (here the sign is the fitness of the particle.)

Figure 1 illustrates the dynamics of the sign-based strategy. At $t = 0$, 1200 particles with random fitness are randomly placed on a 60×60 grid. When moving through the grid, the particles leave signs that represent their current fitness. When deciding the destination node, the particles prefer the cells with signs that represent better fitness values. Using this local rule, the particles tend to form clusters as time progresses. As seen in Fig. 1, the clusters are highly dynamic, dramatically changing in a few generations. The sign-based strategy yields dynamic topologies, but, at the same time, provides a kind of order to the whole system: emergence of clusters of particles that, if isolated, function as a kind of sub-population. Please note that the structure emerges from the local rules, without central coordination. One of the objectives of this paper is to investigate if these structures between order and randomness provided by stigmergic strategies are advantageous when compared to the simple random dynamics of the Brownian strategy.

The following section describes the experiments with the Brownian (B) and sign-based (S) structures. The algorithms are referred in the remaining of the paper has PSO-B and PSO-S, respectively.

Table 1. Pseudo-code of the PSO with sign-based dynamic topology.

PSO-S

---

1. For each particle i: $1 \rightarrow n$:

    1.1. Initialize particle i

    1.2. Evaluate position $\overrightarrow{x_i}$: $f(\overrightarrow{x_i})$

    1.3. Set $p_g(i) = p_i(i) = f(\overrightarrow{x_i})$

2. Set grid size: $q \times s$

3. Place the particles randomly on the grid

4. For each particle i: $1 \rightarrow n$

    4.1. If the fitness of the best position found so far p_j by any of the particles j in the L5 neighborhood of particle i is better than $p_g(i)$, then $p_g(i) = p_j$

    4.2. Place signal (fitness value) on the node.

    4.3. Move if the set of free adjacent nodes is not empty.

        4.3.1. If there are free nodes with signs, move to the node with the sign closer to the optimum fitness value.

        4.3.2. Else, move to a randomly selected free node.

5. For each particle i: $1 \rightarrow n$

    5.1. Update velocity and position.

    5.2. Evaluate position $\overrightarrow{x_i}$: $f(\overrightarrow{x_i})$

    5.2. If $f(\overrightarrow{x_i}) < f(p_i(i))$, then $p_i(i) = \overrightarrow{x_i}$

6. If stop criterion not met, go to 4

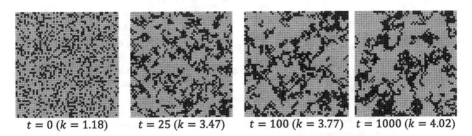

$t = 0 \ (k = 1.18)$ $t = 25 \ (k = 3.47)$ $t = 100 \ (k = 3.77)$ $t = 1000 \ (k = 4.02)$

Fig. 1. Positions of particles at different time-steps t. $q \times s$: 60×60; $\mu = 1200$.

4 Experiments and Results

4.1 PSO-B: Precision and Convergence Speed

The first objective of this study is to compare PSO-B with standard topologies: *lbest*, *gbest* and regular lattice with von Neumann (L5) neighborhood. For that purpose, an experimental setup was constructed with eleven benchmark unimodal and multimodal minimization problems that are commonly used for evaluating the performance of PSO. The functions are described in Table 2: f_1 and f_2 are the unimodal Sphere and Quadric function (also known as Schwefel 1.2 problem); f_3 is the Rosenbrock function, which has one global minimum situated in a narrow, parabolic valley and can be treated as a multimodal problem; f_4–f_9 are multimodal problems with many local optima; f_{10} is the shifted Quadric (f_2) function with noise and f_{11} is the rotated Griewank (f_5) function. The global optimum vector for f_{10} and the orthogonal matrix for f_{11} were taken from the CEC2005 benchmark.

Table 2. Benchmark problems.

Function	Mathematical representation	Range of search/range of initialization	Stop criterion
Sphere f_1	$f_1(\vec{x}) = \sum_{i=1}^{D} x_i^2$	$(-100, 100)^D$ $(50, 100)^D$	0.01
Quadric f_2	$f_2(\vec{x}) = \sum_{i=1}^{D} \left(\sum_{j=1}^{i} x_j \right)^2$	$(-100, 100)^D$ $(50, 100)^D$	0.01
Rosenbrock f_3	$f_3(\vec{x}) = \sum_{i=1}^{D-1} \left(100\left(x_{i+1} - x_i^2\right)\right)^2 + (x_i - 1)^2$	$(-100, 100)^D$ $(15, 30)^D$	10
Rastrigin f_4	$f_4(\vec{x}) = \sum_{i=1}^{D} \left(x_i^2 - 10\cos(2\pi x_i) + 10 \right)$	$(-10, 10)^D$ $(2.56, 5.12)^D$	100
Griewank f_5	$f_5(\vec{x}) = 1 + \frac{1}{4000}\sum_{i=1}^{D} x_i^2 - \prod_{i=1}^{D} \cos\left(\frac{x_i}{\sqrt{i}}\right)$	$(-600, 600)^D$ $(300, 600)^D$	0.05
Schaffer f_6	$f_6(\vec{x}) = 0.5 + \frac{\left(\sin\sqrt{x^2+y^2}\right)^2 - 0.5}{(1.0 + 0.001(x^2+y^2))^2}$	$(-100, 100)^2$ $(15, 30)^2$	0.00001
Weierstrass f_7	$f_7(\vec{x}) = \sum_{i=1}^{D}\left(\sum_{k=0}^{kmax}\left[a^k\cos\left(2\pi b^k(x_i + 0.5)\right)\right]\right) - D\sum_{k=0}^{kmax}\left[a^k\cos\left(2\pi b^k \cdot 0.5\right)\right],$ $a = 0.5, b = 3, kmax = 20$	$(-0.5, 0.5)^D$ $(-0.5, 0.2)^D$	0.01
Ackley f_8	$f_8(\vec{x}) = -20exp\left(-0.2\sqrt{\frac{1}{D}\sum_{i=1}^{D} x_i^2}\right) - exp\left(\frac{1}{D}\sum_{i=1}^{D}\cos(2\pi x_i)\right) + 20 + e$	$(-32.768, 32.768)^D$ $(2.56, 5.12)^D$	0.01
Schwefel f_9	$f_9(\vec{x}) = 418.9829 \times D - \sum_{i=1}^{D} x_i sin\left(\left\|x_i^{\frac{1}{2}}\right\|\right)$	$(-500, 500)^D$ $(-500, 500)^D$	3000
shifted quadric with noise f_{10}	$f_{10}(\vec{z}) = \sum_{i=1}^{D}\left(\sum_{j=1}^{i} z_j\right)^2 * (1 + 0.4\|N(0.1)\|),$ $\vec{z} = \vec{x} - \vec{o}, \ \vec{o} = [o_1, o_2, ...o_D] : \text{shifted global optimum}$	$(-100, 100)^D$ $(50, 100)^D$	0.01
Rotated Griewank f_{11}	$f_{11}(\vec{z}) = 1 + \frac{1}{4000}\sum_{i=1}^{D} z_i^2 - \prod_{i=1}^{D}\cos\left(\frac{z_i}{\sqrt{i}}\right)$ $\vec{z} = M\vec{x}, \ M: \text{ortoghonal matrix}$	$(-600, 600)^D$ $(300, 600)^D$	0.05

Table 3. Statistical measures of the best fitness values empirical distributions (50 runs): mean, standard deviation, median, minimum and maximum values.

		PSO$_{LBEST}$	PSO$_{GBEST}$	PSO$_{VN}$	PSO-B
f_1	Mean (SD)	4.53e−06 (2.31e−06)	3.80e+03 (5.67e+03)	1.54e−09 (1.45e−09)	**4.23e−10** (3.18e−10)
	Median	3.98e−06	3.28e−16	1.03e−09	3.14e−10
	Min/max	1.16e−06/1.10e−05	**4.23e−20**/2.00e+04	2.13e−10/6.89e−09	3.31e−11/1.38e−09
f_2	Mean (SD)	8.44e−13 (1.43e−12)	1.51e+04 (1.17e+04)	2.50e−22 (8.08e−22)	**1.48e−28** (3.66e−28)
	Median	3.67e−13	1.50e+04	1.23e−23	1.55e−29
	Min/max	1.25e−14/9.21e−12	**00e00**/5.67e+04	1.21e−25/5.02e−21	2.27e−31/1.77e−27
f_3	Mean (SD)	4.90e00 (1.47e+01)	1.33e00 (1.85e00)	1.56e00 (2.67e00)	**9.20e−01** (1.62e00)
	Median	6.10e−02	6.78e−04	9.32e−02	6.12e−02
	Min/max	7.34e−06/8.56e+01	**1.80e−08**/4.36e00	3.55e−06/1.53e+01	5.87e−08/4.20e00
f_4	Mean (SD)	1.10e+02 (1.81e+01)	9.86e+01 (2.84e+01)	6.28e+01 (1.67e+01)	**6.23e+01** (1.88e+01)
	Median	1.09e+02	1.00e+02	6.02e+01	6.07e+01
	Min/max	6.57e+01/1.53e+02	4.98e+01/1.61e+02	**3.38e+01**/1.09e+02	3.48e+01/1.24e+02
f_5	Mean (SD)	**2.95e−04** (2.09e−03)	3.80e+01 (5.80e+01)	5.61e−03 (8.78e−03)	7.73e−03 (9.50e−03)
	Median	1.08e−19	4.18e−02	1.09e−19	7.40e−03
	Min/max	**00e00**/1.48e−02	1.08e−19/1.81e02	**00e00**/3.69e−02	**00e00**/4.41e−02
f_6	Mean (SD)	1.94e−04 (1.37e−03)	1.36e−03 (3.41e−03)	**0.00e00** (0.00e00)	**0.00e00** (0.00e00)
	Median	0.00e00	0.00e00	0.00e00	0.00e00
	Min/max	0.00e00/9.72e−03	0.00e00/9.72e−03	**0.00e00**/00e00	**0.00e00**/00e00
f_7	Mean(SD)	**2.86e−05** (0.00e+00)	4.52e00 (2.52e00)	4.43e−02 (1.69e−01)	4.61e−02 (2.50e−01)
	Median	2.86e−05	4.07e00	2.86e−05	2.86e−05
	Min/max	**2.86e−05**/2.86e−05	5.70e−01/1.18e+01	**2.86e−05**/9.84e−01	**2.86e−05**/1.76e00
f_8	Mean (SD)	1.26e−15 (1.64e−16)	7.38e−01 (1.05e00)	**1.12e−15** (2.24e−16)	**1.12e−15** (2.24e−16)
	Median	1.33e−15	2.00e−15	1.33e−15	1.33e−15
	Min/max	**8.88e−16**/1.33e−15	1.33e−15/3.26e00	**8.88e−16**/1.33e−15	**8.88e−16**/1.33e−15
f_9	Mean (SD)	**3.53e+03** (4.99e+02)	3.26e+03 (6.24e+02)	3.07e+03 (4.92e+02)	**2.80e+03** (6.85e+02)
	Median	3.52e+03	3.33e+03	3.04e+03	2.80e+03
	Min/max	2.25e+03/4.68e+03	1.90e+03/4.72e+03	2.27e+03/4.72e+03	**1.80e+03**/4.36e+03
f_{10}	Mean (SD)	2.33e+02 (1.91e+02)	3.10e+02 (1.05e+03)	7.40e00 (3.64e01)	**6.89e00** (3.41e01)
	Median	1.74e+02	3.52e−05	4.76e−02	7.29e−06
	Min/max	3.41e+01/1.07e+03	**1.54e−08**/5.29e+03	4.87e−04/2.05e+02	3.57e−08/1.81e+02
f_{11}	Mean (SD)	3.34e−04 (1.73e−03)	3.62e+01 (5.47e+01)	**5.27e−03** (7.41e−03)	6.60 e−03 (8.28e−03)
	Median	1.09e−19	2.83e−02	1.08e−19	1.08e−19
	Min/max	**0.00e00**/9.86e−03	1.09e−19/1.81e+02	**0.00e00**/2.46e−02	**0.00e00**/2.70e−02

The dimension of the search space is set to $D = 30$ (except f_6, for which $D = 2$). In order to obtain a square grid for the standard *von Neumann* topology, the population size μ is set to 49, a value that lies within the typical range of PSO population size [6]. Following [16], the acceleration coefficients were set to 1.494 and the inertia weight is 0.729. *Xmax* is defined as usual by the domain's upper limit and *Vmax* = *Xmax*. A total of 50 runs for each experiment are conducted. Asymmetrical initialization is used (the initialization range for each function is in Table 2). PSO-B grid size was set to 10×10 and lifespan L of the signs to 1 (please see Sect. 3).

Two experiments were conducted. Firstly, the algorithms were run for a limited amount of function evaluations (49000 for f_1 and f_5, 980000 for the remaining) and the fitness of the best solution found was averaged over 50 runs. In the second experiment

Table 4. Statistical measures of the empirical distributions of evaluations required to meet the stop criteria (50 runs) and success rates (SR).

		PSO$_{LBEST}$	SR	PSO$_{GBEST}$	SR	PSO$_{VN}$	SR	PSO-B	SR
f_1	Mean (SD)	32489.0 (921.5)	50	**16082.4** (2697.4)	33	23530.8 (954.7)	50	22700.7 (906.4)	50
	Median	32364.5		15582		23471		22760.5	
	Min/max	30576/34692		**12642**/24451		21658/25921		20874/24696	
f_2	Mean (SD)	362086.5 (23302.4)	50	125758.5 (22692.6)	6	213141.2 (16261.0)	50	**180237.7** (**11115.8**)	50
	Median	360787		124435.5		214130		180418	
	Min/max	320558/432768		95158/156310		180859/242354		**151851**/209769	
f_3	Mean (SD)	262951.0 (196868.2)	50	**199609.3** (158500.2)	50	392370.0 (196716.6)	49	422088.9 (177503.7)	50
	Median	210381.5		170324		388570		446880	
	Min/max	50274/843927		**16807**/804335		32291/807569		31948/813253	
f_4	Mean (SD)	233260.2 (281453.6)	17	**9602.0** (3599.0)	25	18424.0 (11082.8)	49	15114.5 (3939.7)	48
	Median	77518		8722		15582		14724.5	
	Min/max	21462/86617		**5292**/21805		9604/74872		9359/30968	
f_5	Mean (SD)	30200.7 (1703.9)	50	**14856.1** (2028.1)	27	22015.7 (1304.6)	50	21574.7 (1107.6)	50
	Median	29988		14700		22074.5		21364	
	Min/max	27587/34398		11270/20188		19404/25823		19600/23814	
f_6	Mean (SD)	26263.0 (27266.9)	49	13933.1 (21576.6)	43	17622.4 (16056.7)	50	**10741.8** (10658.2)	50
	Median	18865		6174		12323.5		7570.5	
	Min/max	5243/145334		**1960**/86485		3626/80213		3871/65660	
f_7	Mean (SD)	62057.5 (3031.3)	50	-	0	41677.8 (1360.4)	44	**38976.0** (1654.1)	42
	Median	61201		-		41821.5		38832.5	
	Min/max	56497/69923		-		37730/44296		36652/45766	
f_8	Mean (SD)	35323.1 (1655.2)	50	**17474.6** (2575.5)	32	24420.6 (958.8)	50	24410.9 (1533.3)	50
	Median	35206.5		17297		24206		24083.5	
	Min/max	31556/39249		**13573**/23275		22834/28028		21119/29057	
f_9	Mean (SD)	24883.8 (12417.7)	6	**6394.5** (2626.5)	20	21384.1 (20450.2)	22	12943.9 (3422.2)	31
	Median	23299,5		6125		15288		12054	
	Min/max	9996/46256		**2352**/12593		7693/97069		7693/19894	
f_{10}	Mean (SD)	-	0	661395.9 (136614.7)	40	875793.3 (70624.8)	9	**630285.0** (**119539.5**)	**48**
	Median	-		647510.5		883911		617620.5	
	Min/max	-		**412335**/942760		758961/976962		453642/93428	
f_{11}	Mean (SD)	30302.6 (1670.7)	50	**14112.0** (2006.4)	32	22078.4 (1136.1)	50	21222.9 (995.9)	50
	Median	30282		13622		22001		21070	
	Min/max	26411/33222		**11515**/18767		19355/25088		19306/24549	

the algorithms were run for 980000 evaluations (corresponding to 20000 iterations of standard PSO with $n = 49$) or until reaching the stop criterion. For each problem and each algorithm, the number of evaluations required to meet the criteria was recorded and averaged over 50 runs. The success measure is defined as the number of runs in which an algorithm attains the fitness established as stop criterion.

The results are in Tables 3 and 4. In general, PSO-B is more precise and faster than the other algorithms. Please note that even if *gbest* is faster in several functions, the success rates are significantly lower than those of PSO-B, as expected (see Sect. 2).

The non-parametric Mann-Whitney U test was performed to compare the empirical distributions of fitness values of PSO_{VN} and PSO-B in each function. Results of the tests are significant at $p \leq 0.05$ for f_1, f_2, i.e., the null hypothesis that the two samples come from the same population is rejected. For the remaining functions the null hypothesis is not rejected. As for the speed of convergence, the results of the Mann-Whitney U test comparing PSO_{VN} and PSO-B is significant in functions f_1, f_2, f_4, f_7, f_9, f_{10} and f_{11}.

4.2 PSO-B and PSO-S

Following the comparison with standard topologies, PSO-B has been tested against PSO-S. Results are in Table 5. PSO-B and PSO-S attain equivalent results in most of the problems. However, PSO-S average best fitness is better in f_1 and the result of the statistical test comparing the two distributions is significant.

As for the number of evaluations, PSO-S is fasrer than PSO-B in f_5, f_8 and f_{11} and the result of the statistical test is significant in each case. For f_7, the null hypothesis is not rejected, but PSO-S attains higher success rate (47, against 42 by PSO-B). PSO-S is competitive with PSO-B, being better in three functions. Therefore, we may conclude that a sign-based strategy is the best choice for this test set.

4.3 Grid Size

In the previous experiments, PSO-B and PSO-S were tested on a grid of size 10×10. This value was chosen in order to have a ratio between occupied nodes and total nodes of approximately 1:2. With this ratio, the expected connectivity degree after a random distribution of the particles is roughly $k \sim 3$, lower than $k = 5$ of PSO_{VN} and identical to the degree of *lbest*. The objective was to attain good solutions, like *lbest*, expecting that the dynamics of the connectivity – which makes it possible for a particle to communicate with many other particles in a few generations – would increase convergence speed (*lbest* is precise but slow, while *gbest*, with higher k, is fast but inaccurate). Experiments confirmed this assumption.

Figure 2 describes the variation of PSO-B connectivity degree in four typical runs with different grid size. As predicted, the connectivity degree of grid 10×10 fluctuates around $k = 3$. Smaller grids constrain the movements and increase the expected k value (see k of grid 8×8). The connectivity of 20×20 grids oscillates between 1 and 2, i.e., isolated particles ($k = 1$) occur frequently. As a consequence, the speed of convergence is expected to decrease: large grids create sparse networks in which the flow of information is delayed due to the isolation of the particles.

Table 5. Numerical results and statistical measures of PSO-B and PSO-S.

		PSO-B	PSO-S	PSO-B	PSO-S	PSO-B	PSO-S
		Best fitness		Function evaluations		SR	SR
f_1	Mean (SD)	4.23e−10 (3.18e−10)	**2.48e−10** (2.21e-10)	22700.7 (906.4)	**22491.98 (859.69)**	50	50
	Median	3.14e−10	1.90e−10	22760.5	22442		
	Min/max	3.31e−11/1.38e−09	**1.49e−11**/1.13e−09	20874/24696	**20335**/24353		
f_2	Mean(SD)	1.48e−28 (3.66e−28)	**1.02e−28** (5.35e−28)	180237.7 (11115.8)	**176352.0** (17653.2)	50	50
	Median	1.55e−29	3.41e−30	180418	177894.5		
	Min/max	2.27e−31/1.77e−27	**4.56e-33**/3.78e−27	151851/209769	**148617**/256858		
f_3	Mean (SD)	**9.20e−01** (1.62e00)	2.73e00 (1.04e+01)	422088.9 (177503.7)	**351711.8** (181910.8)	50	50
	Median	6.12e−02	3.16e−02	446880	357430.5		
	Min/max	**5.87e−08**/4.20e00	3.36e−06/7.33e+01	31948/813253	33467/753424		
f_4	Mean (SD)	**6.23e+01** (1.88e+01)	6.55e+01 (1.57+01)	15114.5 (3939.7)	**15109.4** (4194.9)	48	48
	Median	6.07e+01	6.37e+01	14724.5	14455		
	Min/max	**3.48e+01**/1.24e+02	4.28e+01/1.12e+02	9359/30968	**8869**/30233		
f_5	Mean (SD)	7.73e−03 (9.50e−03)	**6.10e−03 (8.81e−03)**	21574.7 (1107.6)	**20814.2 (1465.4)**	50	50
	Median	7.40e−03	1.084e−19	21364	20604.5		
	Min/max	**00e00**/4.41e−02	**0.00e00**/4.65e−02	19600/23814	18473/26509		
f_6	Mean (SD)	**0.00e00** (0.00e00)	**0.00e00** (0.00e00)	**10741.8** (10658.2)	13500.5 (2386.1)	50	50
	Median	0.00e00	0.00e00	7570.5	7644		
	Min/max	**0.00e00**/00e00	**0.00e00**/00e00	3871/65660	**3528**/76734		
f_7	Mean (SD)	4.61e−02 (2.50e−01)	**4.12e−02** (2.21e−01)	**38976.0** (1654.1)	39179.2 (2386.1)	42	47
	Median	2.86e−05	2.86e−05	38832.5	38857		
	Min/max	2.86e−05/1.76e00	2.86e−05/1.50e00	36652/45766	**36015**/49931		
f_8	Mean (SD)	**1.12e−15** (2.24e−16)	1.16e−15 (2.18e−16)	24410.9 (1533.3)	**22864.4** (1092.2)	50	50
	Median	1.33e−15	1.33e−15	24083.5	22907.5		
	Min/max	**8.88e−16**/1.33e−15	**8.88e−16**/1.33e−15	21119/29057	**20384**/25235		
f_9	Mean (SD)	2.80e+03 (6.85e+02)	**2.73e+03** (7.06e+02)	**12943.9** (3422.2)	13063.1 (7845.7)	31	32
	Median	2.80e+03	2.71e+03	12054	10804.5		
	Min/max	1.80e+03/4.36e+03	1.22e+03/4.17e+03	7693/19894	**5684**/50617		
f_{10}	Mean (SD)	**6.89e00** (3.41e+01)	1.33e+01 (5.53e+01)	**630285.0** (119539.5)	657859.4 (118062.2)	48	47
	Median	7.29e−06	1.57e−05	617620.5	636363		
	Min/max	**3.57e−08**/1.81e+02	7.24e−08/3.07e+02	453642/93428	**433650**/918064		
f_{11}	Mean(SD)	6.60 e−03 (8.28e−03)	**6.10e−03** (8.818e−03)	21222.9 (995.9)	**20814.2** (1465.4)	50	50
	Median	1.08e−19	1.08e−19	21070	20604.5		
	Min/max	**0.00e00**/2.70e−02	**0.00e00**/4.65−02	19306/24549	**18473**/26509		

Figure 3 shows the number of particles by their averaged connectivity over a typical run of PSO-B with grid 10×10. Most of the particles have $k = 2$ and $k = 3$. Please remember that the connectivity of *lbest* is $k = 3$. However, as demonstrated in

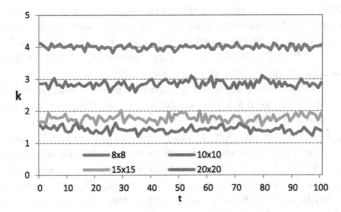

Fig. 2. PSO-B connectivity degree (k) in 100 iterations. Brownian movement.

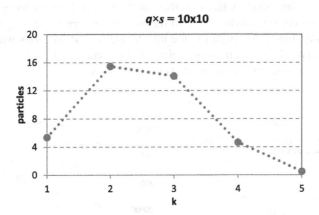

Fig. 3. Number of particles by connectivity degree. Values averaged over 1000 iterations of a typical run. Brownian movement.

the previous section, PSO-B with size 10×10 performs better than *lbest* in the majority of the scenarios. It is plausible that the efficiency of the proposed strategy comes from a combination of factors, namely, the average connectivity degree and the dynamic topology. However, further experiments are required in order to understand better the role of each feature in the performance of the algorithm. Understanding the weight of these factors may lead to more efficient dynamic structures, based on the interaction of the particles and their particular status (fitness, velocity and position).

The significant variation of the average k with the grid is expected to have some kind of impact in the dynamic behavior of the algorithms. The following experiment intends to shed light on the effects of the grid in the performance of the proposed topology. PSO-B and PSO-S were tested with grid size 8×8, 10×10, 15×15 and

20×20 under the same experimental setup described in Sect. 4.1. The average best solution and number of evaluations to meet the stop criteria were averaged over 50 runs and then plotted for each algorithm and each function. Figures 4, 5, 6 and 7 show that the performance of PSO-B and PSO-S may vary with the size of the grid. The nature of that variation depends on the type of fitness landscape.

For unimodal functions (Fig. 4), the performance tends to deteriorate with the size of the grid. In this case, an 8×8 grid would be the best design choice. However, as demonstrated above, PSO-B and PSO-S with grid 10×10 significantly improve other algorithms performance in functions f_1 and f_2. For the multimodal functions, the precision of the solutions tend to improve with grid size, while the convergence speed decreases (see Fig. 5). A 10×10 grid seems to provide a good compromise between speed and accuracy when optimizing f_4, f_5, f_7, f_8 and f_9. Figure 6 shows an atypical behavior of PSO-B and PSO-S on f_3 and f_6. Further studies are required to confirm this hypothesis, but it is possible that this is related to the exceptionality of the functions within the test set: f_3 is a unimodal function that can be treated as multimodal. However, it is different from multimodal functions f_4–f_9, which have many local optima: f_3 has a global minimum situated in a narrow valley that can be very hard to find. Problem f_6 is defined in 2 dimensions, while the other functions have $D = 30$. Independently of the causes for observed behavior, 10×10 seems to be an acceptable design choice for f_3 and f_6. Finally, Fig. 7 shows that in the rotated and shifted functions with noise the algorithms behave like in multimodal problems: the accuracy of the solutions tend to improve with grid size; the convergence speed decreases.

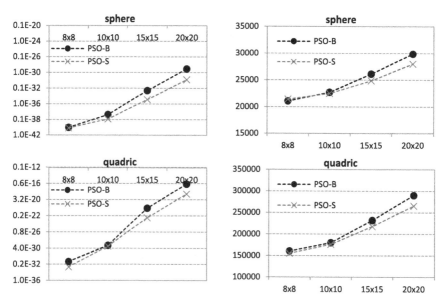

Fig. 4. Sphere (f_1) and Quadric (f_2) functions. Average best fitness (left) and average evaluations (right). PSO-B and PSO-S with different grid size.

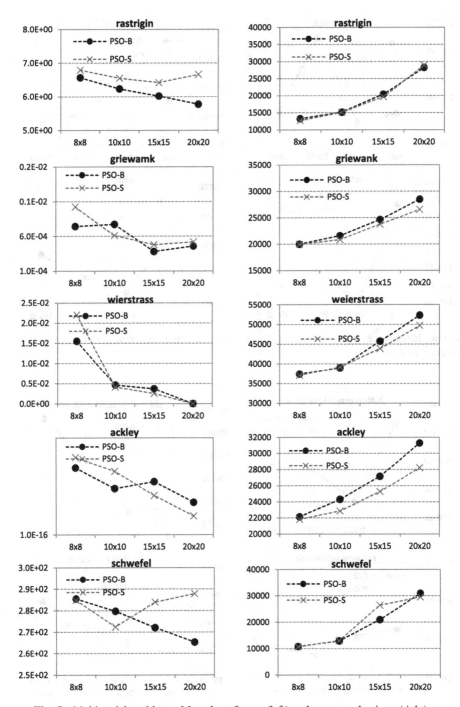

Fig. 5. Multimodal problems. Mean best fitness (left) and mean evaluations (right).

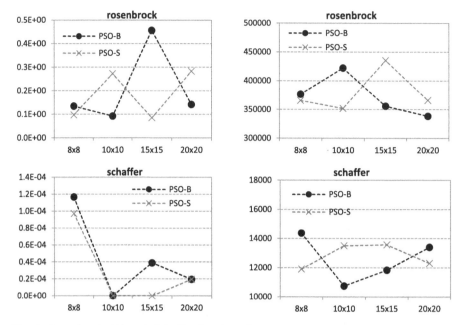

Fig. 6. Rosenbrock (f_3) and Schaffer function (f_6). Average best fitness (left) and average function evaluations.

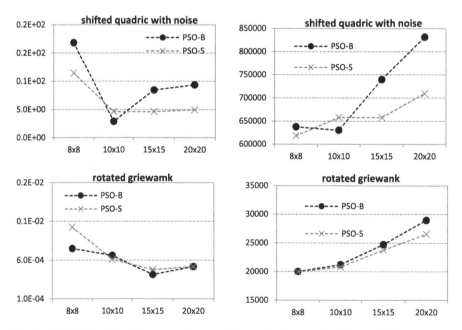

Fig. 7. Shifted Quadric with noise and rotated Griewank. Average best fitness (left) and average function evaluations.

5 Conclusions

This paper investigates a model of dynamic population structures for population-based metaheuristics. Here, the model is tested with the Particle Swarm Optimization (PSO) algorithm. The particles are placed on a 2-dimensional regular network with size $q \times s \geq \mu$, where μ is the population size of the algorithm. The particles move on the grid according to simple rules. The motion can be random (Brownian) or guided by signs that the particle leave on the nodes (stigmergic, or sign-based). The flow of information is defined by the particle's position on the grid and its neighborhood (*von Neumann* vicinity is considered here).

Brownian and sign-based strategies have been implemented, tested and compared to PSOs with standard topologies. The results of the experiments show that the proposed model performs more consistently throughout the test set, improving the other topologies in the majority of the scenarios and under different performance evaluation criteria. The size $q \times s$ of the network affects the performance of the algorithm. However, a grid size that guaranties a ratio $\mu : (q \times s)$ of approximately 1:2 is a good design choice independently of the test function.

Future research will be mainly focused on sign-based movement strategies. Scalability with problem dimension will also be studied. Furthermore, the environmental grid will be investigated as a possible medium for the particles to exchange information about the search.

Acknowledgements. The first author wishes to thank FCT, *Ministério da Ciência e Tecnologia*, his Research Fellowship SFRH/BPD/66876/2009). This work was supported by FCT PROJECT [PEst-OE/EEI/LA0009/2013], EPHEMECH (TIN2014-56494-C4-3-P, Spanish Ministry of Economy and Competitivity), PROY-PP2015-06 (Plan Propio 2015 UGR), and project CEI2015-MP-V17 of the Microprojects program 2015 from CEI BioTIC Granada.

References

1. Augusto, J.P., Nicolau, A.S., Schirru, R.: PSO with dynamic topology and random keys method applied to nuclear reactor reload. Prog. Nucl. Energy **83**, 191–196 (2015)
2. Fernandes, C.M., Laredo, J.L.J., Merelo, J.J., Cotta, C., Nogueras, R., Rosa, A.C.: Performance and scalability of particle swarms with dynamic and partially connected grid topologies. In: Proceedings of the 5th International Joint Conference on Computational Intelligence (IJCCI 2013), pp. 47–55 (2013)
3. Grassé, P.-P.: La reconstrucion du nid et les coordinations interindividuelles chez bellicositermes et cubitermes sp. La théorie de la stigmergie: Essai d'interpretation du comportement des termites constructeurs, Insectes Sociaux, 6, 41–80 (1959)
4. Hseigh, S.-T., Sun, T.-Y., Liu, C.-C., Tsai, S.-J.: Efficient population utilization strategy for particle swarm optimizers. IEEE Trans. Syst. Man Cybern. Part B **39**(2), 444–456 (2009)
5. Kennedy, J., Eberhart, R.: Particle swarm optimization. In: Proceedings of IEEE International Conference on Neural Networks, vol. 4, pp. 1942–1948 (1995)
6. Kennedy, J., Eberhart, R.: Swarm Intelligence. Morgan Kaufmann, San Francisco (2001)

7. Kennedy, J., Mendes, R.: Population structure and particle swarm performance. In: Proceedings of the IEEE World Congress on Evolutionary Computation, pp. 1671–1676 (2002)
8. Morters, P., Peres, Y.: Brownian Motion. Cambridge Press, Cambridge (2010)
9. Ni, Q., Cao, C., Yin, X.: A new dynamic probabilistic particle swarm optimization with dynamic random population topology. In: 2014 IEEE Congress on Evolutionary Computation, pp. 1321–1327 (2014)
10. Liang, J.J., Qin, A.K., Suganthan, P.N., Baskar, S.: Comprehensive learning particle swarm optimizer for global optimization of multimodal functions. IEEE Trans. Evol. Comput. **10**(3), 281–296 (2006)
11. Parsopoulos, K.E., Vrahatis, M.N.: UPSO: a unified particle swarm optimization scheme. In: Proceedings of the International Conference of Computational Methods in Sciences and Engineering (ICCMSE 2004), Lecture Series on Computer and Computational Sciences, vol. 1, pp. 868–887 (2004)
12. Parsopoulos, K.E., Vrahatis, M.N.: Unified particle swarm optimization in dynamic environments. In: Rothlauf, F., et al. (eds.) EvoWorkshops 2005. LNCS, vol. 3449, pp. 590–599. Springer, Heidelberg (2005)
13. Peram, T., Veeramachaneni, K., Mohan, C.K.: Fitness-distance-ratio based particle swarm optimization. In: Proceedings of Swarm Intelligence Symposium, pp. 174–181 (2003)
14. Crepinsek, M., Liu, S.-H., Mernik, M.: Exploration and exploitation in evolutionary algorithms: a survey. ACM Comput. Surv. **45**(3), 35 (2013). article n. 35
15. Shi, Y., Eberhart, R.C.: A Modified particle swarm optimizer. In: Proceedings of IEEE 1998 International Conference on Evolutionary Computation, pp. 69–73. IEEE Press (1998)
16. Trelea, I.C.: The particle swarm optimization algorithm: convergence analysis and parameter selection. Inf. Process. Lett. **85**, 317–325 (2003)

Evaluative Study of PSO/Snake Hybrid Algorithm and Gradient Path Labeling for Calculating Solar Differential Rotation

Ehsan Shahamatnia[1,2(✉)], André Mora[1,2], Ivan Dorotovič[1,3],
Rita A. Ribeiro[1,2], and José M. Fonseca[1,2]

[1] Computational Intelligence Group of CTS/UNINOVA, Caparica, Portugal
{ehs,atm,id,rar,jmf}@uninova.pt
[2] FCT/NOVA University of Lisbon, 2829-516 Monte de Caparica, Portugal
[3] Slovak Central Observatory, Hurbanovo, Slovak Republic

Abstract. PSO/Snake hybrid algorithm is a merge of particle swarm optimization (PSO), a successful population based optimization technique, and the Snake model, a specialized image processing algorithm. In the PSO/Snake hybrid algorithm each particle in the population represents only a portion of the solution and the population, as a whole, will converge to the final complete solution. In this model there is a one-to-one relation between Snake model snaxels and PSO particles with the PSO's kinematics being modified accordingly to the snake model dynamics. This paper provides an evaluative study on the performance of the customized PSO/Snake algorithm in solving a real-world problem from astrophysics domain and comparing the results with Gradient Path Labeling (GPL) image segmentation algorithm. The GPL algorithm segments the image into regions according to its intensity from where the relevant ones can be selected based on their features. A specific type of solar features called coronal bright points have been tracked in a series of solar images using both algorithms and the solar differential rotation is calculated accordingly. The final results are compared with those already reported in the literature.

Keywords: Particle swarm optimization · Snake model · PSO/Snake hybrid algorithm · Gradient path labeling · Image processing · Image segmentation · Object tracking · Solar images

1 Introduction

Particle swarm optimization (PSO), first introduced by [1], is a collective, anarchic, iterative method, with the emphasis on cooperation [2]. PSO is a general search method that can be used to solve a wide range of problems, but it is particularly useful for solving difficult problems, as there are often specific methods for solving easier problems, which are more effective. One such example is the resolution of linear systems, where PSO is not at all the best tool [2]. PSO is a stochastic algorithm based on the analogy of collective behavior of birds' swarms. PSO consists of a population of particles, each similar to a bird searching for the best place to find food. Each particle in PSO is a candidate solution. In PSO, particles are governed under their cognitive and

© Springer-Verlag Berlin Heidelberg 2016
N.T. Nguyen et al. (Eds.): TCCI XXIV, LNCS 9770, pp. 19–39, 2016.
DOI: 10.1007/978-3-662-53525-7_2

social behaviors, which make them able to exchange information and share their experience of explored space, and finally converge towards the optimum of search space, which is the solution to the formulated problem.

One of the main problems in digital image processing is image segmentation, for which more than a thousand different algorithms have been developed [3]. Deformable models are popular spatial segmentation techniques for outlining object boundaries using contours [4]. Active Contour Model (ACM), a technique based on deformable model, was introduced by Kass et al. [5] for 2D image segmentation. The basic idea of ACM, also known as snake model, is to evolve a contour (a curve or a surface) under some constraints to match certain image features. Snake model has been successfully employed in a variety of problem domains such as object tracking, shape modeling image segmentation and stereo vision [5–10].

To drive the snake control points we can use PSO, although limiting the particle search space to avoid premature convergence to the global optimum. There are already several solutions published, for instance Tseng et al. [11] and Li et al. [12] defined multiple PSO populations, in which each control point is confined to a sub-swarm spatially distinct from other sub-swarms. A polar coordinate system that limits the snake control points search space was proposed by Ballerini [13] and Nebti and Meshoul [14]. Another alternative solution proposed by Zeng and Zhou [15] was to iteratively rank the best particles position and by analytical calculations prevent particles from intersecting.

Most of the aforementioned methods act only as a general problem solver and take the approach of formulating the snake model calculations as a minimization problem and then just solving this optimization problem. In this paper, we take the hybrid PSO/Snake approach introduced in [16] and show its versatility by further extending it to solve a real world problem from the astrophysics domain. The PSO/Snake algorithm has already been successfully tested for detection and tracking of small deformable structures such as endothelium cells from cornea microscopic images [17] and tracking sunspots [18].

Snakes model experiences several problems, namely, snake initialization, concave boundaries, sensitivity to noise and local minima. In this work we present a method to customizes PSO to overcomes these problems, maintaining its simple structure. The solution has a low order of complexity and consequently a fast processing time, while precisely calculating the differential rotation of solar features.

In previous paper [19] we presented the PSO/Snake hybrid algorithm for tracking CBPs and calculating solar differential rotation. In this article the PSO/Snake approach is compared with Gradient Path Labeling (GPL) segmentation method, proposed by Mora et al. [20], associated with a region matching process to identify the relevant solar features. The GPL segmentation method uses the image gradient as the basis for a pixel labeling procedure which groups ascending paths that belong to the same regional maximum. Its segmentation result is comparable to the Watershed Transform, with the advantage of having a lower over-segmentation effect, good computation efficiency and customizable segmentation effect. The method produces an image segmented in several intensity regions that are then filtered to match the relevant solar features.

In this paper we address the problem of calculating solar differential rotation to evaluate the PSO/Snake algorithm performance in comparison with the GPL method.

We calculate the solar differential rotation as a function of coronal latitude, by tracking some solar feature as markers. We use a dataset of consecutive images taken by the AIA instrument on board the SDO spacecraft. As solar feature markers for tracking we use Coronal Bright Points (CBPs) which are small and bright structures observed in the extreme ultraviolet (EUV) and the X-ray part of the solar spectrum [21]. They are known to have a mean lifetime of about 8 h. CBPs can reach up to 2×10^8 Km^2 in size but still they look like a tiny shape on the solar images. CBPs are associated with bipolar magnetic features and a large quantity of them (several thousands) emerge over the surface of the Sun per day and thereby in total they bring up huge magnetic fluxes. Physicists and space weather scientists will benefit from a CBP tracking system, since these automatic tools allow them to precisely process large amounts of solar data and consequently improve their solar models [19].

The aim of this paper is to compare the result of applying the PSO/Snake and GPL algorithms for tracking coronal bright points. The tracking results are used for calculating the coronal differential rotation and are cross-referenced with pertinent results reported in the literature.

The remainder of this article is presented in the following order: Sect. 2 looks at the PSO/Snake hybrid algorithms and underlying concepts. The GPL algorithm is described in Sect. 3. The results and discussions are provided in Sect. 4 and in Sect. 5 conclusions and future work are presented.

2 PSO/Snake Hybrid Algorithm

We considered three main reasons to apply the customized bio-inspired PSO/Snake algorithm to the problem of tracking CBPs and calculating solar differential rotation:

- It can overcome the problem of noisy data [16]. CBPs are small in size and the solar images that are used to study them have a fair amount of pixel-size structures that make detecting CBPs more difficult. A noise-tolerant algorithm can improve the accuracy of the detection.
- For precise detection of CBPs, while finding and tracking their shape boundaries, several potential local-optima can be encountered in the process. Bio-inspired solutions, in general, are well suited for multimodal problems [22].
- PSO/Snake algorithm uses PSO dynamics. It is an adaptive algorithm, and can be applied in real world problems of dynamic and uncertain natures. CBPs, similar to sunspots [23], are deformable features on the solar disk that evolve during their life span and because of their loose definition, measuring their evolution bears inherent uncertainty.

As mentioned before, two main concepts driving the hybrid PSO/Snake algorithm are snake model, with its parametric contour concept and PSO with its particle movement mechanism. Snake model is driven by an energy minimization concept. It comprises of an energy function which should be minimized in order to find the optimal contour (snake). The function considers the similarity between the contour and the image features (e.g. object boundaries, image gradient, image intensity, texture, color, etc.) as well as the similarity of the contour to a prior model contour [3].

User interaction can also be involved to define the regions of interest for the snake algorithm [5]. After that whereabouts of the Region Of Interest (ROI) is approximated, the snake will evolve to latch to the exact boundary object edges. Snake model is in essence an optimization algorithm. Canonical snake model proposed by Kass et al. [5] implements the minimization process by solving Euler-Lagrange models of the problem. Details of the canonical snake model and a survey on snake model is provided in [24].

In our model, contour or snake has an energy associated with it, which correlates with the location of the snake in the image and its geometrical characteristics. The idea is to minimize the integral measure which represents the total snake energy, by evolving the snake over time. The original algorithm that controls the snake model iteratively solves a pair of Euler equations on the discrete grid, seeking the minimization of the total snake energy [10]. One of drawbacks is that it is time consuming to achieve good snake models.

To represent snakes we used a parametric approach, since it is computationally efficient and easy to interact with users [25]. Here, the snake is defined as a curve p (s) = (x(s), y(s)), composed by arc lengths where s is the snaxel point. As it is shown in Eq. (1), a number of discrete points called control points or snaxels characterize the snake [5]. In our implementation we use the parametric snake (curve) presentation as follows:

$$p(s,t) = (x(s,t), y(s,t)), \quad s \in [0,1] \tag{1}$$

In this equation, time step for each snaxel point s is denotes by t. Equation (2) shows how the total snake energy is calculated as the sum of integrals of the snake internal energy and external energy coming from the image. In the PSO/Snake hybrid algorithm, the objective function calculates the total snake energy. Since in this implementation the whole population altogether represents one candidate solution to the problem, the objective is to find the contour with the least total snake energy. The lesser the total snake energy, the better it matches the ROI or moves towards it.

$$E_{snake} = \int_0^1 E_{int}(p(s))ds + \int_0^1 E_{ext}(p(s))ds \tag{2}$$

The snake model is considered to be a controlled continuity spline under the influence of internal and external forces, which induce the snake energy. Internal energy consists of two terms, the first and the second derivatives of the snake with respect to s. First term coerces the spline to act like a membrane and the second term makes the snake act like a thin plate [5]. The external energy determines the snake relationship to the image. It is formulated in a way that its local minima corresponds to the image features of interest. Various external energies can be employed such as image intensity, image gradient, object size or shape. One common definition used for gray-level images is the gradient of Gaussian.

PSO is a class of evolutionary optimization algorithms, based on a population of particles (swarm), where each one is a potential solution to the optimization problem. It is the leading part of proposed PSO/Snake hybrid algorithm. Besides the information related to the optimization problem, which represents his position in the search space, each particle is also characterized with a speed. Iteratively, the particles' positions are recalculated according to their velocity, their best solution and also their neighbors' solutions. Each particle position and corresponding fitness score are stored as their best solution and form the cognitive aspect of particle evolution. The influence of neighbors in the position update process is related to their social behavior, which can be defined with various topologies such as ring, star, Von Neumann and random.

If the particle neighborhood is restricted to a subset of swarm it is called local best PSO (*lbest*), while if the neighborhood equals the whole swarm it is called global best PSO (*gbest*). The proposed PSO/Snake hybrid algorithm uses *lbest* with ring structure and radius of 3. The following equations show the dynamics of the canonical PSO algorithm for updating particle velocity and position:

$$v_i(t+1) = \omega(t)v_i(t) + c_1 r_1 (y_i(t) - x_i(t)) + c_2 r_2 (\hat{y}_i(t) - x_i(t)) \tag{3}$$

$$x_i(t+1) = x_i(t) + v_i(t+1) \tag{4}$$

where $x_i(t)$ and $v_i(t)$ are position and velocity of *i-th* particle at time t, $y_i(t)$ and $\hat{y}_i(t)$ denote the best solutions discovered by the i-th particle and its neighborhood up to the time t, i.e. *pbest* and *lbest* respectively. $\omega(t)$ is the inertia weight which controls the impact of the previous velocity and prevents radical changes. Usually, inertia weight is decreased dynamically during the iterative process to balance between exploration in the initial iterations and exploitation when converging to a good solution. Coefficients r_1 and r_2 are uniform random numbers in the range [0,1] to introduce stochastic movement to the PSO particles. Weights of cognitive and social aspects of the algorithm are represented by acceleration factors c_1 and c_2 respectively. As it is shown in [26] regulated values for inertia and acceleration weights can be used to achieve guaranteed convergence. Pseudo-code of a typical standard PSO algorithm is presented in Table 1.

Figure 1 shows the block diagram for a typical scenario where PSO/Snake technique is used for object detection and tracking. In [27] it is discussed how a modular design for solar image processing applications could boost the extendibility and reusability of the applications. As the block diagram shows, PSO/Snake model is implemented in a modular way, making it possible to be customized for specific applications. The PSO/Snake hybrid algorithm integrates the snake model mechanisms with PSO dynamics. While most of swarm intelligence approaches in the literature used in conjunction with snake model tries to optimize the snake model equations, PSO/Snake hybrid does not employ PSO algorithm only as a general problem solver to optimize snake energy minimization, but it also customizes the standard PSO to better solve this specific type of image processing problems. Early experiments on medical image segmentation [16] and sunspot tracking [18] reported promising results. The hybrid model helps to overcome the major drawbacks of traditional snakes: initialization and poor convergence to the boundary concavities, while benefitting from PSO robustness and simplicity. In the Hybrid PSO/Snake model we use a population of particles where each particle is a snaxel of the contour. All particles together form the contour and hence the population itself is the final solution. As the algorithm runs, each particle updates its position and its velocity according to its personal best experience, local best experience, and also according to the internal force of the snake and external force of the image. This gives the PSO/Snake dynamics a wider range of informative guides to update the particle position so that it converges to the ROI.

PSO/Snake hybrid explores the search space according to PSO trajectory disciplines. This eliminates the need to have a separate searching window around each particle as many swarm based snake optimization algorithms do [11, 14, 25]. These methods consider a searching window around each particle and evaluate every position inside that window to determine the snaxels' next position. Since this local search is performed for each particle per iteration, it is a computationally expensive operation that is avoided in the PSO/hybrid model. The velocity update equation in PSO/Snake is as follows:

Table 1. Particle Swarm Optimization pseudo-code

Local-best PSO algorithm
1: Initialize_all_particles( );
2:
3: Do
4:      For each particle
5:          Calculate fitness value
6:          If the fitness value is better than the best fitness value (*pbest*) achieved so far
7:              Set current value as the new *pbest*
8:      End For
9:
10:      For each particle
11:          Set *lbest* as the particle with the best fitness value in the defined neighborhood
12:          Calculate particle velocity
13:          Update particle position
14:      End For
15: While maximum iterations or minimum error criteria is not attained.
16:

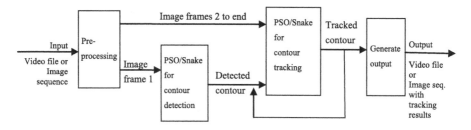

Fig. 1. Block diagram of PSO/Snake algorithm for object detection and tracking

$$v_i(t+1) = \omega v_i(t)$$
$$+ c_1 r_1 (pbest_i(t) - x_i(t))$$
$$+ c_1 r_1 (pbest_i(t) - x_i(t)) \qquad (5)$$
$$+ c_3 r_3 (\bar{x}(t) - x_i(t))$$
$$+ c_4 (f.Image_i)$$

where $pbest_i(t)$ and $lbest_i(t)$ are personal best velocity and local best velocity terms respectively. $\bar{x}(t)$ is the average of positions at time step t, approximating the center of mass of the particles. This term pushes the snake to contract or expand with respect to the sign of its weighting factor, r_3. This term speeds up the algorithm and is particularly useful when the snake is stagnated and there is no other compelling force. If the snake is initialized far from the ROI, this term allows the snake to either expand or shrink towards the ROI and hence it increases the convergence rate and speed. $f.Image_i$ is the normalized image force corresponding to external energy from snake model principles. For particle i, $f.Image_i$ gives the image force at the position specified by that particle. Image gradient or gradient of Gaussian functional are generally accepted for obtaining the image force with acceptable performance. Note that it does not vary along time is computed one single time. c_4 is the weighting factor to control the effect of image force. Inertia weight, ω, is taken to be a relatively small constant and r_1, r_2 and r_3 denote random numbers. Coefficients c_1, c_2, c_3 and c_4 are determined dynamically in a way that, if there is a higher image force c_4 it always gets a higher value. This ensures that if a snaxel is next to the object boundary, it will latch to the object of interest. As Fig. 2 shows, this is implemented by a negative logarithmic function. For each pixel that a particle visits, the dynamic coefficients get negative logarithmic value for the corresponding image force in that pixel. Table 2 provides pseudo-code of the PSO/Snake algorithm used for CBP detection and tracking. A detailed description of the PSO/Snake algorithm is given in [19].

A main difference between PSO and PSO/Snake algorithm is that in the classical PSO population evolves over time, but in the end only one particle (or a limited subset of particles) embody the final solution, where in the PSO/Snake each particle of the population contributes to the solution and final solution is comprised of all particles of the population. In order to control the particle evolution in a tractable manner, velocity control strategies such as velocity strapping is used. PSO being an stochastic approach will provide different results in each because of random seed. PSO/Snake also has stochastic component, but since the initial particle positions are not random, and also because the velocity strapping mechanism along with cognitive and social components of the velocity update prevents particles from drastic changes, the result of several runs of PSO/Snake algorithm is the same. The randomness of the algorithm affects the speed (no of iterations) that it takes to converge to the final results.

Table 2. PSO/Snake pseudo-code

PSO/Snake algorithm for CBP detection and tracking
1:   // Preparation of input images, i.e. normalization of contrast, resolution, sorting, etc.
2:   Input_Image_Prep ( );
3:
4:   // The region of interest for detecting CBP is selected
5:   Select_ROI ( );
6:
7:   // Based on the selected ROI x_i and v_i vectors are initialized
8:   Best_contour ← Initial_Contour ( );
9:
10:
11:   // The weight parameters C_1, C_2, C_3, C_4, w, and random numbers r_1, r_2, r_3 are initialized.
12:   Parameters_Initialization ( );
13:
14:   For each input image
15:       // Calculate $f.Image_i$ with an operator appropriate to the problem, e.g. a GoG function
16:       Calculate_Image_Force ( );
17:
18:       Do
19:           For each particle
20:               Current_contour ← Best_contour;
21:
22:               // Set the best velocity snaxel experienced so far
23:               Calculate_pbest ( );
24:
25:               // Set *lbest* as average of velocities of neighboring particles
26:               Calcualte_lbest ( );
27:
28:               // Update the snaxel velocity and position
29:               Move_Snaxels ( );
30:           End For
31:
32:               Best_contour ← the whole population of snaxels with the lowest energy;
33:       While convergence criteria is not attained.
34:
35:           Calculate_CBP_Specs ( );
36:   End For

3   GPL Algorithm

The Gradient Path Labeling (GPL) segmentation algorithm was designed and proposed to segment retinal images [20] and its accuracy and flexibility made it suitable to be applied in other domains, as it was the case of 2D ion mobility spectra analysis [28] and

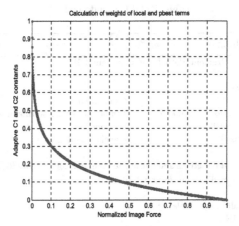

Fig. 2. Dynamic coefficients of C_1, C_2 and C_3. These coefficients control the cognitive, social and expansion behavior of the PSO/Snake population with a negative logarithmic functional of image force. Horizontal axe shows the normalized image force and the vertical axe shows the corresponding value for the dynamic coefficients for a pixel with the corresponding image force.

microscopy image analysis [29]. The segmentation and tracking of CBPs in solar images is also a promising domain for the application of GPL, since CBPs are higher intensity regions with distinguishable boundaries. The GPL is a pixel-level segmentation algorithm that groups ascending paths belonging to the same regional maximum resulting in a segmented image where higher intensity regions are labeled individually. The results are comparable to the Watershed Transform although with better noise independence. Another advantage is that it follows a simple pixel labeling approach allowing it to get a fast segmentation with a complexity proportional only to the image size.

The approach to segment CBPs follows a three step process that starts by preprocessing the image in order to reduce noise, followed by the GPL segmentation. Finally, the generated segmentation regions are filtered to select the region that matches a CBP, and its center of mass location is determined. Table 3 presents the pseudo-code for the GPL algorithm.

3.1 Image Preprocessing

In order to get a more accurate segmentation, a preprocessing step is applied to the original image. The first step is to define the CBP initial position and create a squared ROI centered on this location with a predefined width that encompasses the CBP boundaries to limit the complexity of the GPL segmentation filtering process. Then the ROI, as shown in Fig. 3, is filtered by a 3×3 median filter and green channel is selected for the further processing steps (Fig. 4a-c). As it can be seen in Fig. 4b, the median filter is successful in smoothening the image and removing the salt and pepper noise. It should be noted that for this study we have used the JPEG images (accessible from the SDO image repository from NASA) to test the algorithms. However, for the

Table 3. GPL pseudo-code

Gradient Path Labelling

```
1:    label_I = function Gradient_Path_Labelling(original_I) {
2:
3:      // Initial configuration
4:        curlab = 1;                              // current label
5:        label_I = zeros(size of original_I);     // empty matrix for storing labels
6:        regions = new_list();                    // regions list
7:        equiv = new_list();                      // label equivalences list
8:
9:      // Get Sobel gradient directions and normalize them to 8-connectivity (G_norm)
10:       G_norm = get_norm_directions (original_I);
11:
12:     // Iteration over all image pixels
13:       for all p ∈ label_I {
14:           if (label_I(p) is empty) {           // propagate only from empty labels
15:               label_I(p) = curlab;             // set new label
16:               p_next = p + G_norm(p);          // get next pixel coordinates
17:
18:               // label propagation
19:               while ((p_next is inside boundaries) and (label_I(p_next) is empty)) {
20:                   label_I(p_next) = curlab;    // set to the current label
21:                   p_next = p_next + G_norm(p); // get next pixel coordinates
22:               }
23:
24:               // stop condition: outside the image boundaries or ended on the same label
25:               if (p_next is not inside boundaries or (label_I(p_next) == curlab))
26:                   regions.add(p_next);         // add a new regional maximum
27:
28:           else                                 // ended on a different label
29:                   equiv.add(curlab, label_I (p_next)); // define them as compatibles
30:
31:               curlab = curlab + 1;             // set next label value
32:           }
33:       }
34:
35:     label_I = apply_equivalences(label_I, equiv);   // get final labelled image
36:     label_I = merge_labels(label_I, regions)        // label merging stage
37:     }
38:
39:
40:
```

astrophysical applications the use of the FITS file format, that preserves the raw observation data, is common. In the AIA images used in this study, solar feature locations and boundaries are clearer in the green channel; in the blue channel only the more intense events can be perceived; and in the red channel high intensity CBPs location can usually be observed but less intense ones cannot be perceived (Fig. 3). To obtain the CBP boundaries using the GPL segmentation, it should be applied to the gradient image using the Sobel operator. Comparing the three segmentation results (see

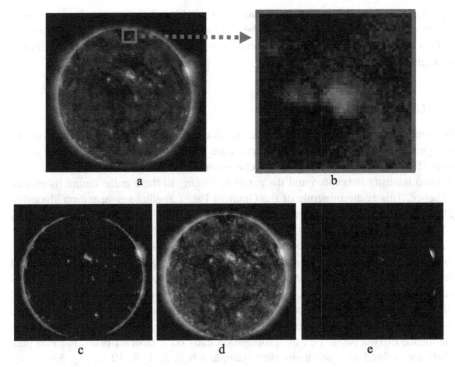

Fig. 3. Solar image of AIA 94Å. (a) Original image with a CBP marked, (b) a zoomed view of the CBP, (c) Red channel of the original RGB-color image, (d) Green channel, (e) Blue channel. (Image courtesy of NASAS/DO) (Color figure online)

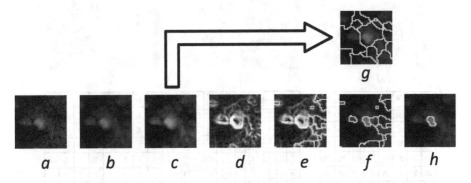

Fig. 4. Example of a CBP detection. (a) original image; (b) median filtered image; (c) green channel; (d) gradient image using Sobel operator; (e) GPL segmentation contours over the gradient image; (f) and (g) GPL segmentation contours over the green image respectively with and without gradient preprocessing; (h) final GPL result. (Color figure online)

Fig. 4) one can see that in the upper image (g) CBP is correctly segmented, although the region boundaries do not overlap the CBP boundaries. However, by applying the GPL to the gradient image (f) this problem is resolved and the CBP boundaries match the segmentation boundaries.

3.2   GPL Image Segmentation

By visualizing an image gradient in a quiver plot it can be noticed that in the proximity of higher intensity regions gradient vectors are directed towards their intensity maximum (Fig. 5a). The confluence of several ascending paths reveals the presence of a regional intensity maximum and the pixels belonging to these paths define its area of influence. This is the principle of the Gradient Path Labeling segmentation algorithm [20]. The algorithm is divided into two main stages: label propagation and labels merging.

The labelling process starts by a sequential pixel analysis following a top-left to bottom-right direction. In this sequence each unlabeled pixel will receive a new label (sequential number) that will be propagated to other pixels along the gradient path. This path is composed by azimuths, obtained using the 3×3 Sobel operator (Fig. 5b), that follow an ascending intensity path. This propagation continues until an already labelled or outside image boundaries pixel is found (Fig. 5c and d). Whenever we have the confluence of two paths, i.e., the propagation ends on a labelled pixel, the two path labels are defined as *equivalents* (for example labels 2, 4, 6, 10 in Fig. 5d and e).

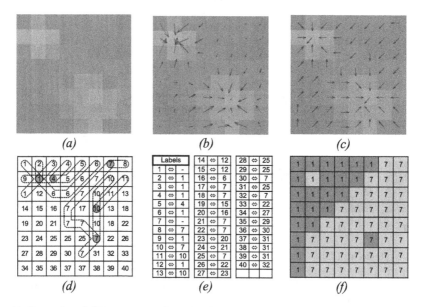

Fig. 5. Example of GPL segmentation algorithm. (*a*) Original image; (*b*) Image gradient; (*c*) Label propagation directions; (*d*) Initial label propagation – (label propagation on the two upper rows); (*e*) Label equivalences table; (*f*) Segmentation results and their maximum point highlighted.

After analyzing all the image, the equivalent labels will be merged together and replaced by the lowest one (Fig. 5f). The GPL result is a segmented image with as many labels as higher intensity regions.

The GPL algorithm has a tendency to produce over-segmented images, in particular when they have flat valleys or flat hills. A merging procedure was introduced to overcome this problem. The merging algorithm is based on a connectivity graph analysis where adjacent regions are merged if they can be connected by a path that does not go lower than a predefined amplitude. At this stage regional maximums are detected and their area of influence is defined.

3.3   CBP Matching

The GPL segmentation produces several regions from which the CBP must be selected. From the segmented regions the most discriminant CBP features are based on color and intensity rather than shape, since CBPs do not have a predefined and static shape. The gradient of the image and the maximum, minimum and mean intensities for each color channel are extracted, as well as the centroid (x, y) location and the difference to its previous location. From a statistical analysis of the extracted features, the most discriminant ones are red and green maximum and mean intensities and centroid location difference. Using these parameters, a multi-criteria score Eq. (6) was computed considering 50 % for location and 50 % for color features. As a result, the highest scored region is selected as the tracked CBP. As it can be seen in the Fig. 4h, GPL is able to detect the CBP boundaries, although at the cost of an over-segmentation effect. By computing the ranking Eq. (6) we are able to identify and select the correct CBP region and track it along consecutive images.

$$
\begin{aligned}
score = {}& 50\% * rank(centroid\,distance)\\
& + 12.5\% * rank(red\,maximum)\\
& + 12.5\% * rank(red\,mean)\\
& + 12.5\% * rank(green\,maximum)\\
& + 12.5\% * rank(green\,mean)
\end{aligned}
\tag{6}
$$

Due to the CBP diversity in intensity and shape this ranking based criteria obtained a better selection accuracy when compared to a relative or absolute one, obtained using computational intelligence techniques such as decision trees. However, whenever the CBP is very fade or simply disappears this selection process will fail to reject all regions. The use of computational intelligence could be applied after the selection process to reject or accept the selected region as a CBP. Since in this work this step was not implemented, regions are selected only if a manual evaluation is available.

4   Results and Discussions

4.1   Dataset and Experiment Setup

As benchmark data we use SDO-AIA corona images, downloaded from NASA-SDO repository[1]. We have used selected images taken at 9.4 nm wavelength in the timespan between 14 September 2010 and 20 October 2010. For comparison purposes we use the manual CBP marking done by an expert in [30] and the database used in [19]. The tracking process starts by choosing a CBP to track. In this comparative study we the CBP marking data done by an expert as our benchmark data. After choosing a CBP, we run PSO/Snake and GPL algorithms independently on that CBP. Each CBP is chosen in the first image and then tracked automatically until it disappears below a predefined size. GPL uses similar criteria, but the decision about stopping the algorithm is embedded in the CBP matching unit. That is if a CBP cannot be matched with a predefined confidence level, the tracking stops. Determining the exact moment for start and ending the lifetime of a CBP is subjective. In this study, since we are comparing the precision of tracking the CBP movement, we consider the life time of CBP according to the manual data availability. For PSO/Snake algorithm the input images are converted to 8-bit grayscale and image force is calculated by a gradient of Gaussian functional with $\sigma = 3$. Images are resized to 512×512 pixels. Altogether, in GPL algorithm we have tested 41 CBP structures, being tracked in 6 days (3098 measurements). PSO/Snake parameters and running conditions are presented in Table 4.

A screenshot of PSO/Snake CBP tracking is shown in Fig. 6. The red circle shows the initial snake around a CBP chosen by an operator. After detection, each CBP is characterized by the heliographic coordinates their center of mass. On the next frame the previous CBP contour is used as a baseline to automatically track its new position. In Fig. 7 is presented the snake contour evolution for a tracked CBP. It shows that, due to the dynamic nature of PSO/Snake hybrid algorithm, detected contours are flexible and can adjust to the dynamic shape and size of deformable objects like CBPs. Figure 8 shows an example of GPL segmentation and tracking of a CBP region in five consecutive images.

4.2   Comparison of Results

First we compare the precision of the GPL and PSO/Snake algorithm in tracking CBP movements against the ground truth obtained by manual tracking data. The calculated angular rotational velocity of solar corona is compared against reported values in the literature to test the feasibility of using these methods in solving real world problems. In the manual procedure [30], an expert operator determined the CBPs positions as they moved. In our study we use the manual CBP markings as our benchmark data.

Several parameters that were reported in the original reference papers [30] and [19] were used for comparison purposes in this paper. Reference [30] manually marks CBPs for the entire dataset, while in [19] PSO/Snake algorithm is used for automatic CBP

---

[1] http://sdo.gsfc.nasa.gov/data/aiahmi/browse.php.

Table 4. PSO/Snake parameters and running conditions

Parameter	Value
No. of iterations	750
No. of particles	15
Image force	Gradient of Gaussian
Image force normalization?	Yes
Gradient sigma	3
C_1, C_2, C_3	Dynamic
C_4	1
r_1, r_2, r_3	Random numbers in range (0,1)
w	0.01
Max velocity	1

Fig. 6. Screenshot of the PSO/Snake CBP tracking. (*Left*) initial snake on the first image at time t, (*Center*) the detected CBP at time t, (*Right*) the tracked CBP at time $t + 10$. The boundary fitting snake is shown with cyan contour, expert's marking is shown with red square and the yellow circle is the center of mass for the tracked CBP by PSO/Snake. (Color figure online)

Fig. 7. The evolution of the snake to detect CBP boundaries can be seen.

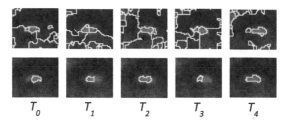

T_0 T_1 T_2 T_3 T_4

Fig. 8. Examples of GPL segmentation and tracking of the CBP region in five consecutive frames.

tracking. Based on the CBP positions, angular rotational velocity of the Sun is calculated. Both references report ω (angular rotation velocity) and $\Delta\omega$ (measurement error at 95 % confidence level). In this study we report measurements for these parameters using the GPL algorithm. Figure 9 shows the calculated rotational speed in different solar latitudes in this study in comparison with the values obtained by other authors. Further details can be found in [30]. The + markers in this image show the tracked CBP structures that, as it can be seen, are well scattered between the ±60° latitude but not in proximity of the solar limb. It is a technically challenging problem to study the solar disk near the limb, since the projection of the sphere shape of the Sun into a 2D image causes projection errors that are even more important in the limbs. For this reason, we focus on comparing the results in the ±5° solar latitude (Fig. 9 Bottom). This figure illustrates the main results comparing the PSO/Snake algorithm vs. GPL algorithm for calculating solar differential rotation. As it can be seen in the figure, both the PSO/Snake and GPL algorithms perform well in approximating the solar rotational velocity in comparison with the ground truth and other reported results. Both methods generate acceptable results matching the scientific community state of the art data. However, PSO/Snake has better conformity with previous results, and a more smooth curve-fitting results, which improves its extendibility in dealing with bigger databases.

To take a closer look at the performance of two methods, results obtained for some sample CBP structures is reported in Table 5. This table shows the comparison between the results obtained with the manual CBP tracking, the results obtained by the PSO/Snake hybrid algorithm and the GPL algorithm for some sample structures. In this table, the structure is the identifier of CBP, b is the heliographic latitude of CBP and ω E is the orbital angular rotation velocity of the Earth that can be looked up from solar almanacs. As with Fig. 9, Table 5, shows that the obtained results in both methods are very close to the results of manual CBP tracking, but with some deviations.

By extending the CBP tracking results for whole life-time of a CBP, we can calculate the accumulated error for the calculated ω by each CBP. Figure 10 show the deviation of the results from the benchmark data for b, $\omega, \Delta\omega$, and ω E, for the PSO/Snake results and the GPL results of all CBP structure during their lifetime. The PSO/Snake algorithm measures the rotational speed of the Sun within ±0.2 of the benchmark data most of the time. For the GPL algorithm, the slight error in calculation of b leads to a greater offset in ω results. The error in this case is −18 degree/day in the

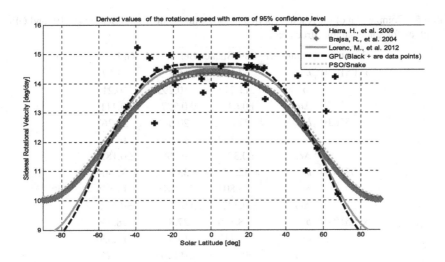

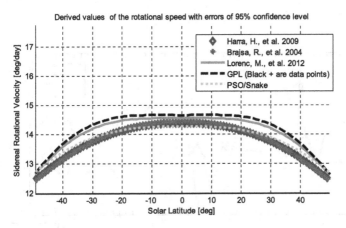

Fig. 9. Rotational speed in different solar latitudes calculated by tracking CBPs in comparison with the values obtained by other works (*Top*). The black dashed curve and the cyan dotted curve show the fit to the mean $\omega(b)$ values as a function of latitude b, for the GPL (present work) and the PSO/Snake algorithms (based on [19]) respectively. Results from [30–32] are superimposed for comparison. The black + markers are the calculated GPL data points (*Bottom*). A cropped and zoomed view of the plot above, confined in the ±50 degree solar latitude. In both images the confidence level is set to 95 %. (Color figure online)

highest. This shows that accumulated deviation error for PSO/Snake algorithm is less than the GPL method. It should be noted that part of this deviation is due to code implementation differences, which, in precise calculations, impose a minute variation. It is also worth mentioning that, in several cases, results displayed bigger differences. Closer investigation by a solar physicist expert (co-author), found out that the PSO/Snake hybrid algorithm behaves consistently and that the user-error is the main cause.

Table 5. Comparison of the results reported in [30] with results from PSO/Snake [19] and GPL algorithms for some sample structures

Structure	Parameter	PSO/Snake	Benchmark	GPL
xy0510.01	*b*	67.103	66.7	67.57
	ω	11.213	10.295	10.21
	$\Delta\omega$	±0.642	±0.327	±0.411
xy0510.03	*b*	21.057	20.5	21.367
	ω	14.387	14.586	14.611
	$\Delta\omega$	±0.303	±0.099	±0.182
xy0510.04	*b*	−32.27	−33.8	−32.409
	ω	13.803	13.648	14.878
	$\Delta\omega$	±0.342	±0.209	±0.432
xy0510.07	*b*	28.170	27.8	28.290
	ω	15.112	14.478	14.530
	$\Delta\omega$	±0.439	±0.116	±0.226

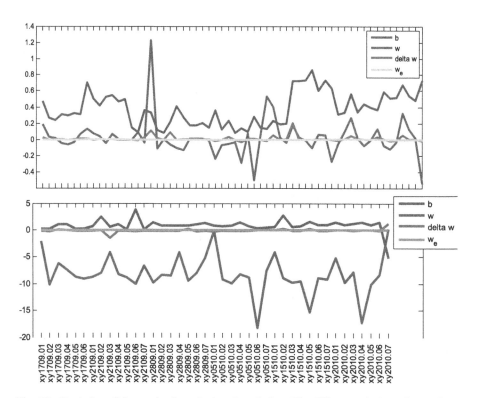

Fig. 10. Deviation of the results from the benchmark data. The differences in *b*, ω, $\Delta\omega$, and ω_E are represented with blue, green, red and cyan lines. (*Top*) The PSO/Snake results error (based on [19]) compared to the benchmark data are within 1.5 degree/day. (*Bottom*) The GPL results error compared to the benchmark data are within 18 degree/day. (Color figure online)

5 Conclusions

In our previous paper [19] we presented the PSO/Snake hybrid algorithm for solving a real solar physic and space weather problem, i.e. tracking CBPs and calculating solar differential rotation. This study compares the PSO/Snake hybrid algorithm and Gradient Path Labeling algorithm. Both methods are applied for determining the speed of sidereal angular rotation of the Sun by analyzing solar images and tracking specific features on a series of successive images. Based on the analysis of the results and comparison with a manual method the resulting values of rotational speed are reliable for both methods. The deviation between these two techniques from the benchmark was similar and both methods calculated solar rotational velocity comparable to the established values reported in the literature. However, by calculating the accumulated deviation during the extended lifetime of all CBP structures, PSO/Snake algorithm demonstrated better performance and lower accumulated deviation. This is because of the dynamic nature of the PSO/Snake algorithm.

For tracking CBP in every new image, information about the CBP profile and its contour is passed from the previous image to the current state, and these data are used only as starting point to recalculate the CBP contour in the current image. The GPL algorithm benefits from the lower computational complexity outperforming the PSO/Snake in calculation speed. PSO/Snake algorithm is an iterative process, which increases its execution time. Particularly calculating image force matrix is the bottleneck of this algorithm. Although GPL achieved reasonable results, PSO/Snake showed a superior performance especially handling the incomplete information and extendibility of the method.

The hybrid combination of PSO with snake model, preserved the PSO simplicity and overcome some of the snake drawbacks. Adding two new terms to the PSO velocity calculation increased the robustness of the algorithm allowing it to evolve even if some component is missing or misleading. The particle/snaxels velocity information embedded in this algorithm, makes it more suitable for object tracking in image processing applications, since it adapts itself to the movement of the object in the images. PSO/Snake algorithm is particularly good in handling noisy data, such as tested solar images. CBPs are small in size and the solar images that are used to study them have a fair amount of pixel-size structures that make detecting CBPs more difficult.

PSO/Snake algorithm is not a general problem solver like PSO. It takes advantage of PSO dynamics, for pushing particle to find the object boundaries in an image. The PSO/Snake algorithm has already been successfully tested for detection and tracking of small deformable structures such as endothelium cells from cornea microscopic images [17] and tracking sunspots [18] and its successful application on CBP test problem can be extended to other problem domains with similar nature. It can be applied in real world problems with dynamic and uncertain nature, where tracking a deformable-shape object is desired.

This work does not consider the splitting and merging of CBPs. As future work we plan to extend the capability of the algorithms for automatic detection of CBPs and analyzing their shape relative to other objects adjacent to them. This will enable the algorithm to determine if a CBP is splitting or merging. A combination of the current

PSO/Snake algorithm for tracking and the GPL capability in CBP matching can be used for implementation of this feature.

Acknowledgments. This work was partially supported by Fundação para a Ciência e a Tecnologia (FCT), MCTES, Portugal through grants SFRH/BPD/44018/2008 (I.D.) and SFRH/BD/62249/2009 (E.S.) and by FCT Strategic Program UID/EEA/00066/203 of UNINOVA, CTS. We would like to also thank the SDO (NASA) and AIA science team for the provided observational material.

References

1. Kennedy, J., Eberhart, R.: Particle swarm optimization. In: IEEE International Conference on Neural Networks, Proceedings, pp. 1942–1948. IEEE, Perth (1995)
2. Clerc, M.: Particle swarm optimization. ISTE Ltd (2006)
3. Morel, J.-M., Solimini, S.: Variational methods in image segmentation: with seven image processing experiments, vol. 14. Springer, Boston (2012)
4. McInerney, T., Terzopoulos, D.: Deformable models in medical image analysis: a survey. Med. Image Anal. **1**, 91–108 (1996)
5. Kass, M., Witkin, A., Terzopoulos, D.: Snakes: active contour models. Int. J. Comput. Vis. **1**, 321–331 (1988)
6. Ballerini, L., Bocchi, L.: Multiple Genetic Snakes for Bone Segmentation. In: Raidl, G.R., Cagnoni, S., Cardalda, J.J., Corne, D.W., Gottlieb, J., Guillot, A., Hart, E., Johnson, C.G., Marchiori, E., Meyer, J.-A., Middendorf, M. (eds.) EvoIASP 2003, EvoWorkshops 2003, EvoSTIM 2003, EvoROB/EvoRobot 2003, EvoCOP 2003, EvoBIO 2003, and EvoMUSART 2003. LNCS, vol. 2611, pp. 346–356. Springer, Heidelberg (2003)
7. Tsechpenakis, G., Rapantzikos, K., Tsapatsoulis, N., Kollias, S.: A snake model for object tracking in natural sequences. **19**, 219–238 (2004)
8. Niu, X.: A Geometric active contour model for highway extraction. In: Proceedings of ASPRS 2006 Annual Conference, Reno, Nevada (2006)
9. Wildenauer, H., Blauensteiner, P., Hanbury, A., Kampel, M.: Motion detection using an improved colour model. In: Bebis, G., et al. (eds.) ISVC 2006. LNCS, vol. 4292, pp. 607–616. Springer, Heidelberg (2006)
10. Karlsson, A., Stråhlén, K., Heyden, A.: A fast snake segmentation method applied to histopathological sections. In: Energy Minimization Methods in Computer Vision and Pattern Recognition. pp. 261–274. Springer Berlin Heidelberg (2003)
11. Tseng, C., Hsieh, J., Jeng, J.: Active contour model via multi-population particle swarm optimization, (2009)
12. Li, R., Guo, Y., Xing, Y., Li, M.: A Novel Multi-Swarm Particle Swarm Optimization algorithm Applied in Active Contour Model. In: Intelligent Systems, 2009. GCIS '09. WRI Global Congress on. pp. 139–143. IEEE (2009)
13. Ballerini, L.: Genetic snakes for medical images segmentation. In: Poli, R., Voigt, H.-M., Cagnoni, S., Corne, D.W., Smith, G.D., Fogarty, T.C. (eds.) EvoIASP 1999 and EuroEcTel 1999. LNCS, vol. 1596, pp. 59–73. Springer, Heidelberg (1999)
14. Nebti, S., Meshoul, S.: Predator prey optimization for snake-based contour detection. Int. J. Intell. Comput. Cybern. **2**, 228–242 (2009)

15. Zeng, D., Zhou, Z.: Invariant topology snakes driven by particle swarm optimizer. In: 2008 3rd International Conference on Innovative Computing Information and Control. p. 38. IEEE (2008)

16. Shahamatnia, E., Ebadzadeh, M.M.: Application of particle swarm optimization and snake model hybrid on medical imaging. In: 2011 IEEE Third International Workshop on Computational Intelligence in Medical Imaging. pp. 1–8. IEEE, Paris, France (2011)

17. Sharif, S.M., Qahwaji, R., Shahamatnia, E., Alzubaidi, R., Ipson, S., Brahma, A.: An efficient intelligent analysis system for confocal corneal endothelium images. Comput. Methods Programs Biomed. **122**, 421–436 (2015)

18. Shahamatnia, E., Dorotovič, I., Ribeiro, R.A., Fonseca, J.M.: Towards an automatic sunspot tracking: Swarm intelligence and snake model hybrid. Acta Futur. **5**, 153–161 (2012)

19. Shahamatnia, E., Dorotovič, I., Fonseca, J.M., Ribeiro, R.A.: An evolutionary computation based algorithm for calculating solar differential rotation by automatic tracking of coronal bright points. J. Sp. Weather Sp. Clim. **6**, A16 (2016)

20. Mora, A.D., Vieira, P.M., Manivannan, A., Fonseca, J.M.: Automated drusen detection in retinal images using analytical modelling algorithms. Biomed. Eng. Online. **10**, 59 (2011)

21. Brajša, R., Wöhl, H., Ruždjak, V., Clette, F., Hochedez, J.-F.: Solar differential rotation determined by tracing coronal bright points in SOHO-EIT images I. Interactive and automatic methods of data reduction. Astron. Astrophys. **374**, 309–315 (2001)

22. Gálvez, A., Iglesias, A.: A new iterative mutually coupled hybrid GA–PSO approach for curve fitting in manufacturing. Appl. Soft Comput. **13**, 1491–1504 (2013)

23. Shahamatnia, E., Dorotovi, I., Mora, A., Fonseca, J., Ribeiro, R.: Data inconsistency in sunspot detection. In: Filev, D., et al. (eds.) Intelligent Systems 2014, pp. 567–577. Springer, Cham (2015)

24. Chen, B., LAI, J.H.: Active contour models on image segmentation: a survey. J. Image Graph. 1, (2007)

25. Horng, M.-H., Liou, R.-J., Wu, J.: Parametric active contour model by using the honey bee mating optimization. Expert Syst. Appl. **37**, 7015–7025 (2010)

26. Van den Bergh, F.: An analysis of particle swarm optimizers, (2002)

27. Shahamatnia, E., Dorotovic, I., Fonseca, J., Ribeiro, R.: On the importance of an automated and modular solar image processing tool. In: Proceedings of the European Planetary Science Congress (EPSC), Portugal (2014)

28. Hakkinen, A., Muthukrishnan, A.-B., Mora, A., Fonseca, J.M., Ribeiro, A.S.: Cell Aging: a tool to study segregation and partitioning in division in cell lineages of Escherichia coli. Bioinformatics **29**, 1708–1709 (2013)

29. Häkkinen, A., Muthukrishnan, A.B., Mora, A., Fonseca, J.M., Ribeiro, A.S.: Cell Aging: a tool to study segregation and partitioning in division in cell lineages of Escherichia coli. Bioinformatics **29**, 1708–1709 (2013)

30. Lorenc, M., Rybanský, M., Dorotovič, I.: On rotation of the solar corona. Sol. Phys. **281**, 611–619 (2012)

31. Hara, H.: Differential rotation rate of X-ray bright points and source region of their magnetic fields. Astrophys. J. **697**, 980 (2009)

32. Brajša, R., Wöhl, H., Vršnak, B., Ruždjak, V., Clette, F., Hochedez, J.-F., Roša, D.: Height correction in the measurement of solar differential rotation determined by coronal bright points. Astron. Astrophys. **414**, 707–715 (2004)

The Uncertainty Quandary: A Study in the Context of the Evolutionary Optimization in Games and Other Uncertain Environments

Juan J. Merelo[1][⊠], Federico Liberatore[1], Antonio Fernández Ares[1], Rubén García[2], Zeineb Chelly[3], Carlos Cotta[4], Nuria Rico[5], Antonio M. Mora[1], Pablo García-Sánchez[1], Alberto Tonda[6], Paloma de las Cuevas[1], and Pedro A. Castillo[1]

[1] Depto. ATC, University of Granada, Granada, Spain
jmerelo@geneura.ugr.es
[2] Escuela de Doctorado, University of Granada, Granada, Spain
[3] Laboratory LARODEC, Institut Supérieur de Gestion, Tunis, Tunisia
[4] Depto. LCC, University of Málaga, Málaga, Spain
[5] Depto. EIO, University of Granada, Granada, Spain
[6] UMR 782 GMPA, Inra, Thiverval-Grignon, France

Abstract. In many optimization processes, the fitness or the considered measure of goodness for the candidate solutions presents *uncertainty*, that is, it yields different values when repeatedly measured, due to the nature of the evaluation process or the solution itself. This happens quite often in the context of computational intelligence in games, when either bots behave stochastically, or the target game possesses intrinsic random elements, but it shows up also in other problems as long as there is some random component. Thus, it is important to examine the statistical behavior of repeated measurements of performance and, more specifically, the statistical distribution that better fits them. This work analyzes four different problems related to computational intelligence in videogames, where Evolutionary Computation methods have been applied, and the evaluation of each individual is performed by playing the game, and compare them to other problem, neural network optimization, where performance is also a statistical variable. In order to find possible patterns in the statistical behavior of the variables, we track the main features of its distributions, *skewness* and *kurtosis*. Contrary to the usual assumption in this kind of problems, we prove that, in general, the values of two features imply that fitness values do not follow a normal distribution; they do present a certain common behavior that changes as evolution proceeds, getting in some cases closer to the standard distribution and in others drifting apart from it. A clear behavior in this case cannot be concluded, other than the fact that the statistical distribution that fitness variables follow is affected by selection in different directions, that parameters vary in a single generation across them, and that, in general, this kind of behavior will have to be taken into account to adequately address uncertainty in fitness in evolutionary algorithms.

© Springer-Verlag Berlin Heidelberg 2016
N.T. Nguyen et al. (Eds.): TCCI XXIV, LNCS 9770, pp. 40–60, 2016.
DOI: 10.1007/978-3-662-53525-7_3

1 Introduction

Optimization methods usually need a single-valued and reliable feedback on the quality of possible solutions to work correctly. This value, usually called *cost* or *fitness*, informs the algorithm on the goodness of the solution and, when facing different alternatives, it is used to select a particular solution over others. This does not imply the necessity of a single floating point number as feedback; since these methodologies are based on comparisons, it is usually enough if the values can be partially ordered. In multiobjective optimization [11], for instance, two solutions can even be considered *non-comparable*, based on the set of fitness values they possess. In either case, the answer to the question "Is this solution better than the other?" needs to be either a 'Yes', or 'No', or 'Not decidable' in most optimization algorithms.

For many problems, however, the fitness or cost of a solution cannot be described by a single value or vector, because there is *uncertainty* when measuring it. Such uncertainty is inherent to most real-world physical systems, such as the one described in [9], where a control system is optimized through a stochastic procedure, but it also shows in machine learning and, in general, when the optimization algorithm uses a lower-level method which is, itself, stochastic. In these cases, the best way to describe the quality of a solution will be a random variable, not a single value or a vector of deterministic values. In our research, we routinely find this phenomenon in different optimization problems, such as:

- Optimizing the layout of a web-page using Simulated Annealing (SA) [48]. Since SA is a stochastic procedure, the fitness of an obtained solution will be a random variable.
- Training any kind of neural network [9,40]; in the first case, also mentioned above, we dealt with a physical installation, introducing another kind of randomness. Since training a neural network is usually a stochastic procedure, the error rate obtained after every training run will also follow a statistical distribution.
- Evolving game bots (autonomous agents) [43]. In this case, the uncertainty arises from the problem itself; in games, several factors such as the initial positions of the players or the opponent's behavior add further stochastic components, so that the final score will also be *uncertain* or *noisy*. In some cases, too, the bot itself will rely on probabilities to generate its behavior [15], in which case two different runs with exactly the same initial conditions and opponent will also yield different scores.
- In coevolutionary algorithms [15,46,47], individuals are evaluated by randomly choosing opponents from a pool, thus resulting in a fitness that is variable in a single generation and across generations [42].

In all these examples, it cannot be said that there is actually *noise* added to a *real* fitness. Instead, the fitness itself can be represented with a statistical variable, whose value arises from a stochastic process, evaluation, or training. In this sense, we are not concerned with the origin of this uncertainty. It could be noise in the measuring process, uncertainty in the fitness itself, or incomplete information like, for instance, what appears when surrogate models are used. It might be the case that the statistical nature of the fitness as a random variable might be different, distribution-wise, but we think that, except in the case that uncertainty is created by adding noise to fitness, the results obtained here will hold. In the problems used here, the randomness arises from the inherent stochasticity of the methods used to measure fitness, which, themselves, have a random element.

Despite a considerable amount of literature on problems with stochastic fitness values, there is a distinct lack of exhaustive research on the behavior of fitness functions, seen as random variables. That is why, after an initial study of noise in a specific game in [38], we dug into experimental data discovering that, even if the distribution in that particular case was always a Gamma, the parameters of the distribution were different. In that study, we proposed a solution to the noise issue, based on using the Wilcoxon comparison [57] as a selection operator. This meant that the random variable behaved in different ways depending on the particular individual, the state of evolution and, of course, the specific problem.

But, more importantly, this initial conclusion disagrees with the usual assumptions of optimization in uncertain environments, where it is usual to take a normal distribution of noise with fixed σ as the initial hypothesis [2]. For instance, in the functions of the Black Box Optimization Benchmarks [26], uncertainty was simulated by adding a Cauchy noise function centered in 0, that is, a centered, sharp bell-shaped distribution, with different widths. Either multiplicative or additive noise has been used in different occasions. However, our initial work hints that this is not the case in real-world optimization problems, ultimately invalidating the generality of the conclusions on different optimization methods obtained through the usual benchmarks. Besides, we also prove in [38] that, depending on the shape of the statistical distribution of the fitness, different methods could yield the best results. While methods that use the median or average would work well in centered distributions, other methods such as our technique, based on the Wilcoxon test, are better in more uncertain environments or, of course, in the case the noise distribution is not centered.

In this paper, we collect data from several different case studies, which will be presented later on, to find a stochastic model for the fitness using statistical tools. Our final objective is to eventually build a model as general as possible, able to account for most sources of uncertainty; failing that, to devise selection operators that are able to work with random fitness in a natural way.

This second part, if needed, will be the focus of future research.

The rest of the paper is organized as follows. In Sect. 2 we present the state of the art for evolutionary algorithms in uncertain environments, to be followed by a short presentation of the four problems with uncertainty whose measures will be used in this paper in Sect. 3. Results will be presented in Sect. 4, followed by our conclusions.

2 State of the Art

The most comprehensive, although not recent, review of the state of the art for evolutionary algorithms in *uncertain* environments is presented by Jin and Branke in [29], while more novel papers such as [3,49,50] include brief updates. Goh and Tan [23] performed a similar survey, focused on multiobjective optimization.

In their survey, Jin and Branke state that uncertainty is categorized into noise, robustness issues, fitness approximation, and time-varying fitness functions. In addition, different options for dealing with the uncertainty are discussed. In principle, the approach presented in this paper is designed to model the first kind of uncertainty, namely, noise or uncertainty in fitness evaluation. It can be argued that there is uncertainty in the true fitness as stated in the third category. However, we think that, in general, the third issue refers to the case in which expensive fitness functions are replaced by surrogate functions which carry a certain amount of error, and whose value varies as the surrogate models are updated. Independently from the origin of uncertainty, Jin and Branke suggest several methods to tackle it, based either on *implicit* / explicit averaging over fitness measures [13,27] or on a threshold imposed during the selection phase. Papers such as Stroud's [56], Esteban-Díaz's [13] or Di Mario's [12] use this kind of approach to deal with noise. Other authors [18] propose to use new rank-based selection and mutation operators in order to evolve a neural network topology used as a controller for a robot. Results show that those operators are suitable for problems where the fitness landscape is noisy, but it is still using a central value for the fitness that might not be always valid.

Since then, several other solutions for uncertainty have been proposed. A usual approach for scientists more focused on obtaining a straightforward solution to the optimization problem without modifying existing tools and methodologies, is just to disregard the noise in the fitness and take whatever value is returned by a single evaluation, often after re-evaluating all individuals at each generation. This option seems to work especially well if the population is large [27], since the selective pressure is lower and solutions have the chance to be evaluated several times before being selected or discarded; this leads, if the population is large enough, to an *implicit averaging* as mentioned in [29]; in fact, Rattray and Shapiro [53] in their theoretical model compute by how much the *crisp* population must be enlarged to overcome the *problem* of noise. This solution is exploited in our previous research in games, although one evaluation in some of these works consists, in fact, of an average of several evaluations, on different maps or considering different opponents, see for instance [34,42,43], or in the evolution of neural networks [4,40].

The key to the efficiency of this approach stems from the fact that selection used in evolutionary algorithms is usually stochastic, so uncertainty in fitness evaluation could have the same effect as randomness in selection or a higher mutation rate, which might make the evolutionary process easier in some particular cases [50]. In fact, Miller and Goldberg proved that an infinite population would not be affected by noise [41] and Jun-Hua and Ming studied the effect of noise in convergence rates [31], proving that an elitist genetic algorithm finds at least one solution in noisy environments with probability one, although with a lowered convergence rate. This possible positive effect of uncertainty in evaluation leads to some authors calling it "a blessing and the curse" in the context of surrogate models [44], which, as we have seen before, carry with them a degree of uncertainty and randomness.

In real-world problems, however, populations are finite: in fact, using large populations decreases the algorithm's efficiency and can be time consuming, so the usual approach for dealing with fitness with a degree of randomness is to enlarge the population to a value bigger than would normally be needed in a non-noisy environment, while keeping it to a manageable size. Furthermore, it has been proved recently that using two parents to generate offspring, that is, crossover, is able to successfully deal with noise [20], while an evolutionary algorithm based mainly on mutation, such as the $\mu+1$ EA, or evolutionary programming [19], would suffer a considerable degradation of performance. However, crossover is part of the standard kit of evolutionary algorithms, so using it and increasing the population size has the advantage that no special provision or change in the implementation has to be made. There is no big decreasing in efficiency as long as oversized populations are not used. Using oversized populations, however, might have a good effect on the algorithm in general, if appropriate computational resources are available [33].

Another way to deal with uncertainty which is more theoretically sound is using *real* averaging, that is, a statistical central tendency indicator, which usually is the *average*; average happens to be equal to the median in the case of the random variable following the normal distribution. In this case, resampling is used to acquire a statistically significant amount of measures and then the average is computed over them. This strategy has been called *explicit averaging* by Jin and Branke, and it is used, for instance, in [31]. Explicit averaging decreases the fitness variance, thus reducing uncertainty, but defining the appropriate sample size for the averaging process is not straightforward [1]; besides, this central tendency might not be representative if the noise is not centrally distributed, as proved in [39]. Our research group uses this approach in some cases, with an important difference: individuals are not re-evaluated every additional generation, but their fitness value is the average of several evaluations, performed immediately [42]. Most authors use several measures of fitness for each new individual [10], although other averaging strategies have also been proposed, for example averaging over the neighbourhood of the individual or using *resampling*, that is, heuristically requiring more fitness measurements [35]. This assumes that there is, effectively, a real average of the fitness values, which is true for Gaussian random noise and other distributions (such as Gamma's or Cauchy's), but it does not necessarily hold for all distributions. In this paper, we are going to model these distributions in order to verify whether this assumption is indeed correct.

To the best of our knowledge, other central tendency measures such as the median, which might be more adequate for certain noise models, have not been tested; the median always exists, while the average might not exist for non-centrally distributed variables. Besides, most models keep the number of evaluations fixed and independent of its value, which might result in bad individuals being evaluated multiple times before finally being discarded; some authors have proposed *resampling*, [51,52], which will effectively increase the number of evaluations and thus slow down the search. In any case, using explicit averaging usually requires just a small change to the algorithm framework, by using the average of several evaluations as the new fitness function. Thus, it is usually the method preferred by researchers and practitioners using off-the-shelf libraries such as ECJ [37].

In order to improve the efficiency of the algorithm, or the running time, these two averaging approaches that are focused on the evaluation process might be complemented with changes to the selection process. For instance, a threshold [52,54] that is related to the noise characteristics to avoid making comparisons of individuals that might, in fact, be very similar or statistically the same; this is usually called *threshold selection* and can be applied either to explicit or implicit averaging fitness functions. Uncertainty can also be used to compare different algorithms, with some authors proposing, instead of taking more measures, testing different solvers [8], some of which might be more affected by noise than others. However, recent papers have proved that sampling might be ineffective [49] in some types of evolutionary algorithms, adding running time without an additional benefit in terms of performance. This is one lead we will try to follow in the current paper, by modeling noise in order to eventually design an algorithm that behaves correctly in that environment.

All the aforementioned approaches still face the issue of the statistical representation of the *true* fitness, even more so if there are instead several measures that represent, *as a set* the fitness of an individual, such as the case study described in [39]. This is what we have been using in many of our papers: a method that uses resampling via a memory attached to every individual that stores all fitness measures and uses either explicit averaging or statistical tests like the non-parametric Wilcoxon test. In order to test this approach on benchmark problems more realistic that the ones adopted so far, we need to characterize the noise that actually appears in games and other real-world case studies for optimization.

3   Case Studies Used in This Paper

The fitness analyzed in the four different case studies, all related to computational intelligence in games, are described in this paper: generation of character backstories in a MAssive Drama Engine for non-player characters (MADE), described in Subsect. 3.1, optimization of bots for playing the real time strategy game (RTS) Planet Wars in Subsect. 3.2, optimization of the ghost team in Ms. Pac-Man, which will be described in Subsect. 3.3, automatic generation of autonomous players for the famous RTS StarCraft, explained in Subsect. 3.4 and an artificial neural network optimization problem using an EA Subsect. 3.5.

These five problems have been chosen for two main reasons: the origin of uncertainty is different for each of the case studies; and data for the experiments is readily available, with the possibility of running further experimental trials, if needed. In the case of MADE, fitness is computed through a simulation; in the case of Planet Wars, the bot themselves have a random component, with its representation including probabilities of different courses of action; in Ms. Pac-Man, uncertainty lies in the nature of the game itself; and the huge amount of possibilities in StarCraft, with a considerable number of units behaving independently, creates an extremely high source of uncertainty. These scenarios are not a complete representation of all possible causes of uncertainty in optimization, but we think that the sample is big and varied enough to generalize the obtained results, which will be presented in the next section.

In all cases, three generations were chosen, and they are different depending on the problem. We feature the first generation (except in the case of MADE), a intermediate generation and one close to the end of the evolution, containing individuals close to the solution. These were chosen to check the progress of the two statistical parameters in different situations: close to random in the case of the first generations, and close to the *real* value, in the case of the last ones. The particular number of generations is not really important, the importance is how close they are to the end of the evolution, which is different in each case.

We will next examine the creation of character backstories in the problem called MADE.

3.1 Creation of Character Backstories

MADE [21] is a framework for the automatic generation of virtual worlds that allow the emergence of backstories for secondary characters that can later on be included in videogames. In this context, an *archetype* is a well-known behaviour present in the imaginary collective (for example, a "hero" or a "villain"). Given a fitness that takes into account the existence of different N_a archetypes for a virtual world, MADE uses a Genetic Algorithm (GA) [24] to optimize the parameter values of a Finite State Machine (FSM) that models the agents of that world. For the evaluation, a world is simulated using this parameter set, and the log is analyzed to detect behaviours of the world agents that match the desired archetypes.

As the evolved parameters are the probabilities to jump from one state to another in the FSM, each fitness evaluation is performed executing the virtual world five times per individual, obtaining the average fitness. Selection is, therefore, performed comparing this average fitness, with a binary tournament. Fitness values range from 0 to N_a, and are calculated taking into account the rate of occurrence of the archetypes in the execution log.

3.2 A 'Simple' Real-Time Strategy Game: Planet Wars

Planet Wars [16] is a simple Real-Time strategy (RTS) game. In RTS games, the objective is to defeat the enemy using resources available in the map to build and manage units and structures: differently from turn-based strategy games, in RTS all choices have to be performed in real time.

Planet Wars provides a simplification of the usual elements in RTS games: one kind of unit (spaceships) and one kind of resources and structures (planets). Spaceships are automatically generated on the planets controlled by a player, and they are used to conquer enemy planets, the main way to defeat the enemy.

In this paper we are using the results obtained from the Genebot algorithm [17]. This algorithm optimizes seven parameters of a hand-coded FSM, two of which are probabilities. These values are used in expressions used by the bot to take decisions, such as the selection of the target planet to attack or reinforce; this implies that the actions of the bot will be different every time the bot acts, that is, some state transitions are based on probabilities. Fitness is calculated confronting the bot obtained from the parameter set of the FSM five times against a competitive hand-coded bot. The result of each match takes into account the 'slope' of the number of player spaceships during the time of the match. Positive results mean that the bot won, as the slope will be positive, and vice versa. Theoretical values are in the range $[-1, 1]$, although these extremes are impossible to attain in the game. A value of -1 would indicate that the player lost all their ships at startup, while 1 would mean the contrary: it generated all the spaceships and won in the initial time. The fitness of an individual is the sum of all five results, and therefore is in the range $[-5, 5]$.

3.3   Ghost Team Optimization

Ms. Pac-Man is a variant of the famous Pac-Man game that extends its mechanics with several extra features, the most interesting being the inclusion of a random event that reverses the direction of the ghosts. This game is used in the Ms. Pac-Man vs Ghosts competition [36], where participants can submit controllers for both Ms. Pac-Man and the Ghost Team, the first trying to maximize its score, the second trying to minimize Ms. Pac-Man's. The framework used to test the methodology analyzed defines the following restrictions for the Ghost Team:

- A ghost can never stop and if it is in a corridor it must move forward.
- A ghost can choose its direction only at a junction.
- Every time a ghost is at a junction the controller has to provide a direction from a set of feasible directions.
- After 4000 game ticks, a level is considered completed and the game moves on to the next one.

In the methodology applied to this case study, published in [34], the fitness of each individual is computed as the maximum score obtained by eight different Ms. Pac-Man controllers. Some of these controllers are the champions of past editions of the international competition, so they are very tough rivals for the ghost team.

3.4   A Complex Real-Time Strategy Game: StarCraft

StarCraft has become a *de facto* testbed for AI research in complex RTS games [45]. In fact, given the high variety of game features, such as configuration options, game modes, units, maps, etc.; along with the existence of several frameworks and tools related with it; researchers have exploited the game for a great variety of topics: micro and macro management of units, temporal and spatial reasoning, battle planning, combat results prediction, optimal paths and dealing with problems such as the one that is the topic of this paper, uncertainty in the evaluation of the fitness, among others.

The individuals described in this subsection have been generated using Star-CraftGP [22], a Genetic Programming (GP) [32] framework that automatically generates the source code of high-level strategies of bots. In this case, Linear GP [55] was used to generate the building order of the units to create, and also the rules to activate during the game: for example, when to attack the enemy or when to collect more materials. Each individual is a source code file in C++ that is compiled during the evaluation. A population of 32 individuals was evolved during 30 generations. The rest of the parameters used are presented in [22].

As in some of the games described above, the fitness of one individual is computed pitting the bot against different enemies, each one following a different strategy. More specifically, in this case every individual faces three *divisions* of enemies (considered as weak, medium and strong rivals), each division containing four different enemies.

The original fitness function assigned a higher weight to a victory against the stronger enemies, i.e. it used a lexicographical fitness, so one victory in a higher tier was considered better than more victories in the immediately lower one. For example, one individual that wins 1 time against one enemy of every tier was considered better than one individual that beats all individuals from the medium and weak tier, but none in the strong one.

Conversely, to ease comparisons among noise in the present study, we have calculated an aggregated fitness function that still respects this decision, that is, prioritizing victories of harder divisions, by giving different weights to each one. The following equation describes the fitness function:

$$F_{StarCraft} = 21 \times A + 5 \times B + C + R \tag{1}$$

where A is the number of victories against enemies in the strongest tier, B is the number of victories against the middle tier, and C is the number of victories against the weakest enemies. Thus, for example, one victory in the middle tier is worth more points (5 points) than 4 victories in the weak tier (4 points). Also, a coefficient of the aggregated score at the end of all the games, R has been added, in order to deal with ties in number of victories. This is in fact an internal score computed by the game, that takes into account all the aspects of a match, ranging from the number of kills to the type and quality of units built.

Moreover, the evaluation process is quite time-consuming, so in order to save execution time, at least one enemy of the weak tiers must be defeated before allowing the individual to proceed to fight the next one. To this end, if a bot does not win against weaker rivals, we consider it cannot defeat the stronger ones: so the evaluation terminates at that point, with the current score.

3.5   Artificial Neural Networks Optimization Using an EA: GProp

The design of an Artificial Neural Network (ANN) [28] requires to set both, the structure of its set of hidden layers, along with the parameters it uses, weights and learning constants).

G-Prop ("genetic backpropagation") [5–7] aims to solve the problem of finding appropriate initial weights, number of neurons in the hidden layer and learning parameters for a Multilayer Perceptron (MLP) with a single hidden layer. It does so by combining an EA and the QuickProp method [14] for training MLPs. The EA selects the MLP's initial weights, picks its learning rate, and changes the number of neurons in the hidden layer through the application of specific genetic operators. Since this representation of the neural network is then trained using QuickProp, which is an stochastic gradient-descent algorithm, the results will have a certain variability, resulting in the uncertainty in the fitness that will be studied here. In this problem, fitness is the classification accuracy or success rate; this fitness is obtained after training the MLP, which sets its weights, and then testing it on a test set. After classification, weights will be different, so the test result will also be, making fitness *noisy* and thus amenable to analysis in the paper.

4   Experiments and Results

With the case studies presented above, data on fitness values is collected by selecting a few random individuals in every generation of the considered EAs, and measuring their fitness 100 times, intentionally using much more repetitions than a normal optimization method would. Thus, every individual is represented by a random variable sampled 100 times. According to the usual assumptions, this random variable should follow a normal distribution, with a certain σ and centered on the *true* fitness value. In order to verify this hypothesis, we plotted the distribution's *skewness*, that is, its asymmetry, and its *kurtosis*, which is a parameter related to its shape [25] A symmetrical distribution, like the normal distribution, has skewness and kurtosis equal to 0; asymmetric distributions, such as the Gamma that we had found in previous papers [38], have non-zero skewness and kurtosis, related to their θ and κ parameters, for instance. These parameters are what defines the statistical distribution; κ is the shape parameter and skewness is $2/\sqrt{\kappa}$, which means that it is only 0, corresponding to normality, if κ grows to infinity. Kurtosis is $6/\kappa$, implying the same. A random variable can have skewness and kurtosis fixed at any value: thus, we present these values in the following figures, with skewness plotted as the x axis against kurtosis on the y axis.

Figure 1 represents skewness and kurtosis in the MADE case study, for which we took measures of a variable amount of individuals every generation, from

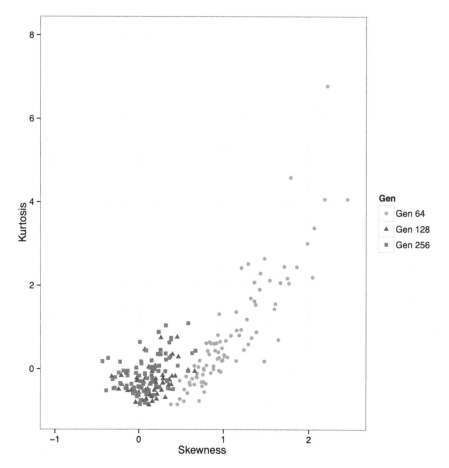

Fig. 1. Skewness and kurtosis for fitness in several generations of the MADE problem. Different colors represent different generations. (Color figure online)

100 in generation 64 to around 50 in the latest generation. The number was variable because some of them stopped before finishing. Anyway, the number of measurements is enough for the statistical analysis. You can already see that the distribution is not normal, since almost no individual has a kurtosis and skewness equal to zero; some of them, however, are close. This will be the case for the rest of the experiments, too; in some very limited cases fitness distribution will be almost normal in the first or the last generations, but that will never be the case for all individuals or even a significant fraction, nullifying the hypothesis of fitness behaving like a crisp fitness with gaussian added noise. As generations proceed, a curious convergence towards the normal distribution is observed; in the first generations, values of skewness and kurtosis are quite high and correspond to an arbitrary distribution (Beta or uniform): however, as the simulation proceeds, the two values approach zero. It must be noted, however, that they do not converge

exactly to 0, meaning that, even if uncertainty in this case can be approached by a normal distribution, such an approximation would only be correct for the latest generations of the simulation. In general, individual fitness in MADE will follow an arbitrary distribution with a general shape and asymmetry.

The shape of the graph for the Planet Wars case study, shown in Fig. 2 for two different generations, is different but shares some similarities. The dispersion also decreases as evolution proceeds, with the shape of the distribution becoming closer to the normal distribution in generation 50. Nevertheless, the initial kurtosis is quite high and values above 2 and below 0 are found even later in the evolution. Noise is, thus, *noisy* and does not conform to a single shape, even less a normal one; this implies that using a single statistical model to represent noise will never be too close to reality, since the shapes of the statistical distribution are, in general, quite different from the normal distribution and then different

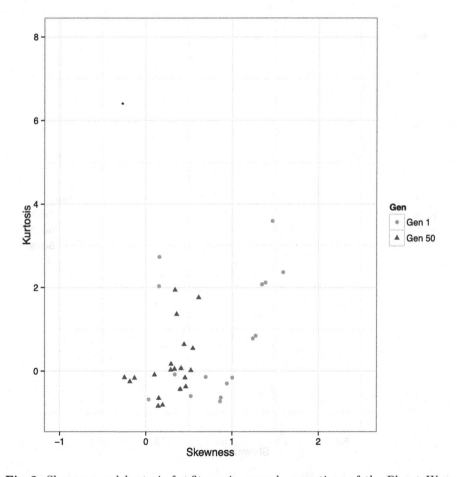

Fig. 2. Skewness and kurtosis for fitness in several generations of the Planet Wars problem. Different colors represent different generations. (Color figure online)

among themselves even for a single problem, that is, the shape of the statistical distribution of fitness values in uncertain environments is, itself, uncertain or *noisy*.

The graph for the third case study, ghosts in Ms. Pac-Man, is different in several aspects, and is shown in Fig. 3. First we have to take into account, as explained in Subsect. 3.3, that differently from the previous cases, the fitness for a ghost team is the maximum, not an average of several values. This causes a curious behavior of fitness: in the first generation, several individuals have *crisp* values; however, this is less and less true, becoming more "random" as generations proceed, that is, the set of values the fitness has got begins to have many different values while in the first generations it had one or a few. To put it in another words, in the first generation the set of fitness measures could look like {x x x y x x x}. As evolution proceeds, the measures in the set tend to be all

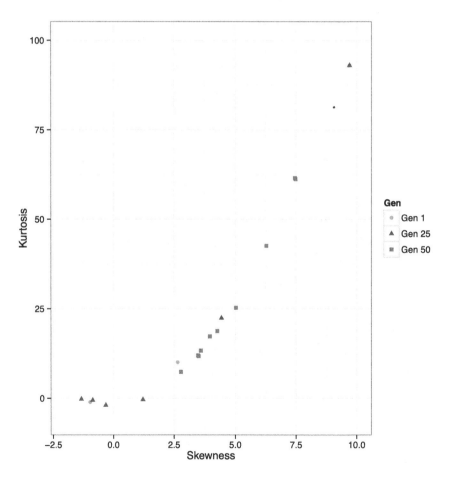

Fig. 3. Skewness and kurtosis for fitness in several generations of the Ms. Pac-Man problem. Different colors represent different generations. (Color figure online)

different That is why the behavior shown in the graph is completely different: distributions get increasingly asymmetric and their shape grows further away from a normal distribution and closer to a Beta distribution. Even if the trend is different from the other two problems, the overall aspect is the same: there is no single distribution that is able to describe the shape of fitness with an uncertainty component, for all considered generations.

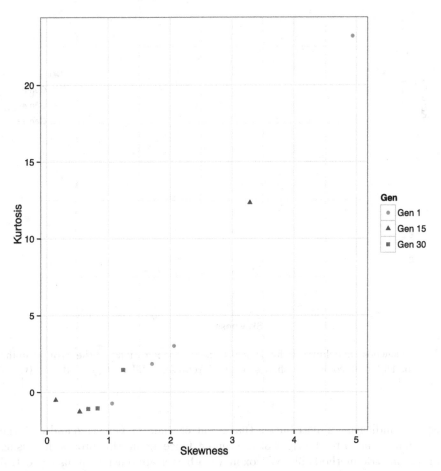

Fig. 4. Skewness and kurtosis for fitness in several generations of the StarCraft game. Different colors (or shades of gray) represent different generations. (Color figure online)

The last game we have evaluated is StarCraft, with kurtosis and skewness shown in Fig. 4. In this case evaluation takes a very long time, that is why only a few samples were available. That might be the reason it is not quite clear if there is a trend. The latest generation seems to be a bit closer to normal distribution, but intermediate generations tend to have a high value. However, even if values seem to be closer in generation 30, they are in some cases positive and in other

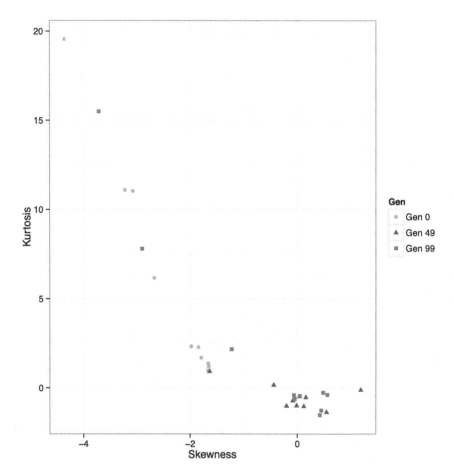

Fig. 5. Skewness and kurtosis for fitness in several generations of the MLP training problem. Different colors (or shades of gray) represent different generations. (Color figure online)

negative, indicating a distribution that is flatter than the Gaussian and with the *bump* more loaded to the right of the center. Once again, this proves that using non-parametric methods like Wilcoxon are a better approach than using central measures such as the average.

For the sake of completeness, we have also included in this paper a problem that comes from a different area: genetic optimization of neural networks. The skewness/kurtosis graph is included in Fig. 5. Since the problem is completely different, the distribution of the values is also completely different. For starters, skewness tends to be negative, indicating distributions with a long tail to the right; that means that, even if the value is centered along a particular value, there are many values that are larger that this central value. Once again, resampling cannot change the fact that the average will not be an accurate description for the whole data. Besides, values tend to get closer to 0 although in every generation there are values quite far away from them; e.g. in the last generation, a neural net whose fitness distribution has kurtosis of 15, indicating a very sharp bump, is present, but it also has a low kurtosis of almost −4 indicating a long tail to the right. The conclusion in this case is similar in the sense that values tend to change while they keep away from a single kurtosis and skewness; even having less values than the latter if both of them are set to 0.

5 Conclusions

In this paper, we set out to study the statistical distribution that best fits the stochastic fitness values of single individuals in several case studies in the area of games; we have also included a genetically optimized neural network for the sake of comparison. Stochastic optimization evolutionary algorithms applied to MADE, Planet Wars, Ms. Pac-Man, and StarCraft exploit different ways to compute the fitness values, but for all of them the fitness value is not a fixed number but a random variable. This is also the case in G-Prop, the genetically optimized multilayer perceptron. We prove the hypothesis that not only noise does not follow the normal, or Gaussian, distribution, or other centrally-distributed models such as Cauchy, which have been used repeatedly in literature for benchmarking selection methods in the presence of noise; but also it does not follow a single, particular distribution even when considering a single case study or a single generation.

This conclusion follows from our study of the parameters of the statistical variables that describe fitness. The best way to describe them is using two parameters: kurtosis and skewness. These two parameters have been computed and plotted for candidate solutions extracted from each of the case studies, proving that not only distributions are asymmetrical and not bell-shaped, but that their shape changes within a single problem and in different stages of the computation; this is in accordance with the conclusions reached by Rattray and Shapiro in [53] for evolution of finite populations in the presence of noise. In some cases, like MADE, it seems clear that due to the fact that averages are used as a representative for selection, individuals whose fitness is closer to a central shape are oversampled and thus selected preferably, with almost-central individuals in the latest stages being a consequence of this fact. In other cases, when fitness is computed in a different way or selection takes another form, the effect is exactly the opposite. Using averages or other central measures like the median does not seem to be supported by the results of this paper since in many cases and almost always in the early stages of the evolution, fitness, being a random variable, does not pass a centrality test and it might not even possess a reliable, that is, statistically significant, average. A better way of comparing any fitness with uncertainty would be, as proposed by the authors, using non-parametric tests such as the Wilcoxon test that impose a partial order on the individuals [38]. This partial order can be used, in several different ways, for selection.

The fact that there is no single model representing the distribution of fitness also implies that it is an error to use centrally distributed random variables added to an actual fitness to test operators and algorithms that operate in uncertainty. Either real values should be used, such as the ones proposed above, or a distribution with varying shape and symmetry such as Beta. However, in this case we should take into account that "true" or "crisp" fitness *does not really exist*, so any modelization of uncertain values that uses noise added to a fitness value is, in the more general case, wrong, although it might still return correct results in some cases. If the fitness evaluation is expensive and tests have to be performed for new selection operators, the best way to model uncertainty would be to use *different* statistical models applied to every individual, with different skewness and kurtosis. However, this would be only a first-order approximation and it might still favor methods that use averages. Following the model proposed by Jin [30] for surrogate models, assuming normality in fitness will make selectable some individuals that should not be. Assessing this error and its impact on selection, and comparing how different methods, such as the one based in statistical techniques and proposed previously, reduce that error is also left as future work.

What remains to be done is to effectively apply Wilcoxon-based comparisons to the case studies above. Since real-world case studies are computationally expensive to evaluate, we plan to create a benchmark for problems with uncertainty which reflects in the best possible way how fitness is organized in a wide array of problems. In order to attain this goal, we will examine as many uncertain problems as possible, in the attempt to deduce a model of noise what as general as possible.

Acknowledgements. This work has been supported in part by projects TIN2014-56494-C4-3-P (Spanish Ministry of Economy and Competitiveness), SPIP2014-01437 (Dirección General de Tráfico), PRY142/14 (Fundación Pública Andaluza Centro de Estudios Andaluces en la IX Convocatoria de Proyectos de Investigación), PROY-PP2015-06 (Plan Propio 2015 UGR), and project CEI2015-MP-V17 of the Microprojects program 2015 from CEI BioTIC Granada. We would like also to thank the anonymous reviewers for this paper, for suggesting new readings and avenues of research.

References

1. Aizawa, A.N., Wah, B.W.: Scheduling of genetic algorithms in a noisy environment. Evol. Comput. **2**(2), 97–122 (1994)
2. Arnold, D.: Evolution strategies in noisy environments-a survey of existing work. In: Kallel, L., Naudts, B., Rogers, A. (eds.) Theoretical Aspects of Evolutionary Computing. Natural Computing Series, pp. 239–250. Springer, Heidelberg (2001). doi:10.1007/978-3-662-04448-3_11
3. Bhattacharya, M., Islam, R., Mahmood, A.: Uncertainty and evolutionary optimization: a novel approach. In: 2014 IEEE 9th Conference on Industrial Electronics and Applications (ICIEA), pp. 988–993, June 2014
4. Castillo, P.A., González, J., Merelo-Guervós, J.J., Prieto, A., Rivas, V., Romero, G.: G-Prop-III: global optimization of multilayer perceptrons using an evolutionary algorithm. In: GECCO 1999: Proceedings of the Genetic and Evolutionary Computation Conference, p. 942 (1999)

5. Castillo, P.A., Merelo-Guervós, J.J., Prieto, A., Rivas, V., Romero, G.: G-Prop: global optimization of multilayer perceptrons using GAs. Neurocomputing **35**, 149–163 (2000). http://dx.doi.org/10.1016/S0925-2312(00)00302-7, available from http://geneura.ugr.es/pub/papers/castilloNC.ps.gz

6. Castillo, P., Carpio, J., Merelo-Guervós, J.J., Rivas, V., Romero, G., Prieto, A.: Evolving multilayer perceptrons. Neural Process. Lett. **12**, 115–127 (2000). http://dx.doi.org/10.1023/A:1009684907680

7. Castillo, P., Merelo-Guervós, J.J., Prieto, A., Rojas, I., Romero, G.: Statistical analysis of the parameters of a neuro-genetic algorithm. IEEE Trans. Neural Netw. **13**(6), 1374–1394 (2002). http://ieeexplore.ieee.org/iel5/72/22620/01058074.pdf

8. Cauwet, M.L., Liu, J., Teytaud, O., et al.: Algorithm portfolios for noisy optimization: compare solvers early. In: Learning and Intelligent Optimization Conference (2014)

9. Chiaberge, M., Merelo, J.J., Reyneri, L.M., Prieto, A., Zocca, L.: A comparison of neural networks, linear controllers, genetic algorithms and simulated annealing for real time control. In: De Facto, B. (ed.)Proceedings of the European Symposium on Artificial Neural Networks, pp. 205–210 (1994). Index available from http://www.dice.ucl.ac.be/esann/proceedings/esann1994/content.htm, available from http://polimage.polito.it/~marcello/articoli/esann.94.jj.pdf, and a scanned version from http://www.dice.ucl.ac.be/Proceedings/esann/esannpdf/es1994-533-S.pdf

10. Costa, A., Vargas, P., Tinós, R.: Using explicit averaging fitness for studying the behaviour of rats in a maze. In: Advances in Artificial Life, ECAL, vol. 12, pp. 940–946 (2013)

11. Deb, K.: Multi-objective Optimization Using Evolutionary Algorithms, vol. 16. John Wiley & Sons, New York (2001)

12. Di Mario, E., Navarro, I., Martinoli, A.: A distributed noise-resistant particle swarm optimization algorithm for high-dimensional multi-robot learning. In: Robotics and Automation (ICRA), pp. 5970–5976, May 2015

13. Esteban-Diaz, J., Handl, J.: Implicit and explicit averaging strategies for simulation-based optimization of a real-world production planning problem. Informatica (03505596) **39**(2) (2015)

14. Fahlman, S.: Faster-learning variations on back-propagation: an empirical study. In: Proceedings of the 1988 Connectionist Models Summer School. Morgan Kaufmann (1988)

15. Fernández-Ares, A., Mora, A.M., García-Arenas, M., Guervós, J.J.M., García-Sánchez, P., Castillo, P.A.: Co-evolutionary optimization of autonomous agents in a real-time strategy game. In: Esparcia-Alcázar, A.I., Mora, A.M. (eds.) EvoApplications 2014. LNCS, vol. 8602, pp. 374–385. Springer, Heidelberg (2014). doi:10.1007/978-3-662-45523-4_31

16. Fernández-Ares, A., Mora, A.M., Guervós, J.J.M., García-Sánchez, P., Fernandes, C.: Optimizing player behavior in a real-time strategy game using evolutionary algorithms. In: IEEE Congress on Evolutionary Computation, pp. 2017–2024. IEEE (2011)

17. Fernández-Ares, A., Mora, A.M., Merelo, J.J., García-Sánchez, P., Fernandes, C.M.: Optimizing strategy parameters in a game bot. In: Cabestany, J., Rojas, I., Joya, G. (eds.) IWANN 2011. LNCS, vol. 6692, pp. 325–332. Springer, Heidelberg (2011). doi:10.1007/978-3-642-21498-1_41

18. Flores, D.: Rank based evolution of real parameters on noisy fitness functions: evolving a robot neurocontroller. In: 10th Mexican International Conference on Artificial Intelligence (MICAI), pp. 72–76. IEEE (2011)

19. Fogel, L.J., Owens, A.J., Walsh, M.J.: Artificial Intelligence Through Simulated Evolution. John Wiley, New York (1966)
20. Friedrich, T., Kötzing, T., Krejca, M., Sutton, A.M.: The Benefit of Sex in Noisy Evolutionary Search. ArXiv e-prints, February 2015
21. García-Ortega, R.H., García-Sánchez, P., Mora, A.M., Merelo, J.: My life as a sim: evolving unique and engaging life stories using virtual worlds. In: ALIFE 2014: The Fourteenth Conference on the Synthesis and Simulation of Living Systems, vol. 14, pp. 580–587 (2014)
22. García-Sánchez, P., Tonda, A.P., Mora, A.M., Squillero, G., Guervós, J.J.M.: Towards automatic starcraft strategy generation using genetic programming. In: 2015 IEEE Conference on Computational Intelligence and Games, CIG 2015, Tainan, Taiwan, 31 August – 2 September 2015, pp. 284–291. IEEE (2015)
23. Goh, C.K., Tan, K.C.: An investigation on noisy environments in evolutionary multiobjective optimization. IEEE Trans. Evol. Comput. 11(3), 354–381 (2007)
24. Goldberg, D.E.: Genetic Algorithms in Search, Optimization and Machine Learning. Addison Wesley, Reading (1989)
25. Groeneveld, R.A., Meeden, G.: Measuring skewness and kurtosis. The Statistician, 391–399 (1984)
26. Hansen, N., Finck, S., Ros, R., Auger, A.: Real-parameter black-box optimization benchmarking 2009: noisy functions definitions (2009)
27. Hansen, N., Niederberger, A.S., Guzzella, L., Koumoutsakos, P.: A method for handling uncertainty in evolutionary optimization with an application to feedback control of combustion. IEEE Trans. Evol. Comput. 13(1), 180–197 (2009)
28. Haykin, S.: Neural Networks: A Comprehensive Foundation, 2nd edn. Prentice Hall PTR, Upper Saddle River (1998)
29. Jin, Y., Branke, J.: Evolutionary optimization in uncertain environments - a survey. IEEE Trans. Evol. Comput. 9(3), 303–317 (2005). Cited by (since 1996) 576
30. Jin, Y.: Surrogate-assisted evolutionary computation: recent advances and future challenges. Swarm Evol. Comput. 1(2), 61–70 (2011)
31. Jun-hua, L., Ming, L.: An analysis on convergence and convergence rate estimateof elitist genetic algorithms in noisy environments. Optik Int. J. Light Electron Opt. 124(24), 6780–6785 (2013). http://www.sciencedirect.com/science/article/pii/S0030402613007730
32. Koza, J.R.: Genetic Programming - on the Programming of Computers by Means of Natural Selection. Complex Adaptive Systems. MIT Press, Cambridge (1993)
33. Jiménez Laredo, J.L., Dorronsoro, B., Fernandes, C., Merelo, J.J., Bouvry, P.: Oversized populations and cooperative selection: dealing with massive resources in parallel infrastructures. In: Nicosia, G., Pardalos, P. (eds.) LION 2013. LNCS, vol. 7997, pp. 444–449. Springer, Heidelberg (2013). doi:10.1007/978-3-642-44973-4_47
34. Liberatore, F., Mora, A., Castillo, P., Merelo, J.: Comparing heterogeneous and homogeneous flocking strategies for the ghost team in the game of Ms. Pac-Man. IEEE Trans. Comput. Intell. AI Games PP(99), 1 (2015)
35. Liu, J., St-Pierre, D.L., Teytaud, O.: A mathematically derived number ofresamplings for noisy optimization. In: Proceedings of the 2014 Conference Companion on Genetic and Evolutionary Computation Companion, GECCO Comp 2014, pp. 61–62. ACM, New York (2014). http://doi.acm.org/10.1145/2598394.2598458
36. Lucas, S.M.: Ms Pac-Man versus ghost-team competition. In: 2009 IEEE Symposium on Computational Intelligence and Games, CIG 2009, p. 1, September 2009

37. Luke, S., Panait, L., Balan, G., Paus, S., Skolicki, Z., Bassett, J., Hubley, R., Chircop, A.: ECJ: a java-based evolutionary computation research system (2006). Downloadable versions and documentation can be found at the following url: http://cs.gmu.edu/eclab/projects/ecj

38. Merelo, J.J., Castillo, P.A., Mora, A., Fernández-Ares, A., Esparcia-Alcázar, A.I., Cotta, C., Rico, N.: Studying and tackling noisy fitness in evolutionary design of game characters. In: Rosa, A., Merelo, J.J., Filipe, J. (eds.) ECTA 2014 - Proceedings of the International Conference on Evolutionary Computation Theory and Applications, pp. 76–85 (2014)

39. Merelo, J.J., Chelly, Z., Mora, A., Fernández-Ares, A., Esparcia-Alcázar, A.I., Cotta, C., Cuevas, P., Rico, N.: A statistical approach to dealing with noisy fitness in evolutionary algorithms. In: Merelo, J.J., Rosa, A., Cadenas, J.M., Dourado, A., Madani, K., Filipe, J. (eds.) Computational Intelligence. SCI, vol. 620, pp. 79–95. Springer, Heidelberg (2016). doi:10.1007/978-3-319-26393-9_6

40. Merelo-Guervós, J.J., Prieto, A., Morán, F.: Optimization of classifiers using genetic algorithms, pp. 91–108. MIT Press (2001). Chap. 4, iSBN:0262162016, draft available from http://geneura.ugr.es/pub/papers/g-lvq-book.ps.gz

41. Miller, B.L., Goldberg, D.E.: Genetic algorithms, selection schemes, and the varying effects of noise. Evol. Comput. **4**(2), 113–131 (1996)

42. Mora, A.M., Fernández-Ares, A., Merelo-Guervós, J.J., García-Sánchez, P., Fernandes, C.M.: Effect of noisy fitness in real-time strategy games player behaviour optimisation using evolutionary algorithms. J. Comput. Sci. Technol. **27**(5), 1007–1023 (2012)

43. Mora, A.M., Montoya, R., Merelo, J.J., Sánchez, P.G., Castillo, P.Á., Laredo, J.L.J., Martínez, A.I., Espacia, A.: Evolving bot AI in unreal^{TM}. In: Chio, C., Cagnoni, S., Cotta, C., Ebner, M., Ekárt, A., Esparcia-Alcazar, A.I., Goh, C.-K., Merelo, J.J., Neri, F., Preuß, M., Togelius, J., Yannakakis, G.N. (eds.) EvoApplications 2010. LNCS, vol. 6024, pp. 171–180. Springer, Heidelberg (2010). doi:10.1007/978-3-642-12239-2_18

44. Ong, Y.S., Zhou, Z., Lim, D.: Curse and blessing of uncertainty in evolutionary algorithm using approximation. In: 2006 IEEE Congress on Evolutionary Computation, CEC 2006, pp. 2928–2935. IEEE (2006)

45. Ontañón, S., Synnaeve, G., Uriarte, A., Richoux, F., Churchill, D., Preuss, M.: A survey of real-time strategy game AI research and competition in starcraft. IEEE Trans. Comput. Intellig. AI Games **5**(4), 293–311 (2013)

46. Paredis, J.: Coevolutionary computation. Artif. Life **2**(4), 355–375 (1995)

47. Parras-Gutierrez, E., Arenas, M.G., Rivas, V.M., del Jesus, M.J.: Coevolutionof lags and RBFNs for time series forecasting: L-Co-R algorithm. Soft Comput. **16**(6), 919–942 (2012). http://dx.doi.org/10.1007/s00500-011-0784-2

48. Peñalver, J.G., Merelo, J.J.: Optimizing web page layout using an annealed genetic algorithm as client-side script. In: Eiben, A.E., Bäck, T., Schoenauer, M., Schwefel, H.-P. (eds.) PPSN 1998. LNCS, vol. 1498, pp. 1018–1027. Springer, Heidelberg (1998). doi:10.1007/BFb0056943. http://www.springerlink.com/link. asp?id=2gqqar9cv3et5nlg

49. Qian, C., Yu, Y., Jin, Y., Zhou, Z.-H.: On the effectiveness of sampling for evolutionary optimization in noisy environments. In: Bartz-Beielstein, T., Branke, J., Filipič, B., Smith, J. (eds.) PPSN 2014. LNCS, vol. 8672, pp. 302–311. Springer, Heidelberg (2014). doi:10.1007/978-3-319-10762-2_30

50. Qian, C., Yu, Y., Zhou, Z.H.: Analyzing evolutionary optimization in noisy environments. CoRR abs/1311.4987 (2013)

51. Rada-Vilela, J., Johnston, M., Zhang, M.: Population statistics for particle swarm optimization: resampling methods in noisy optimization problems. Swarm Evol. Comput. **17**, 37–59 (2014). http://www.sciencedirect.com/science/article/pii/S2210650214000261
52. Rakshit, P., Konar, A., Nagar, A.: Artificial bee colony induced multi-objective optimization in presence of noise. In: 2014 IEEE Congress on Evolutionary Computation (CEC), pp. 3176–3183, July 2014
53. Rattray, M., Shapiro, J.: Noisy fitness evaluation in genetic algorithms and the dynamics of learning, pp. 117–139 (1998)
54. Rudolph, G.: A partial order approach to noisy fitness functions. In: Proceedings of the IEEE Conference on Evolutionary Computation, ICEC, vol. 1, pp. 318–325 (2001)
55. Squillero, G.: MicroGP-an evolutionary assembly program generator. Genet. Program Evolvable Mach. **6**(3), 247–263 (2005). http://dx.doi.org/10.1007/s10710-005-2985-x
56. Stroud, P.D.: Kalman-extended genetic algorithm for search in nonstationary environments with noisy fitness evaluations. IEEE Trans. Evol. Comput. **5**(1), 66–77 (2001)
57. Wilcoxon, F.: Individual comparisons by ranking methods. Biometrics Bull. **1**(6), 80–83 (1945)

Hybrid Single Node Genetic Programming
for Symbolic Regression

Jiří Kubalík[1(✉)], Eduard Alibekov[1,2], Jan Žegklitz[1,2], and Robert Babuška[1,3]

[1] Czech Institute of Informatics, Robotics, and Cybernetics,
CTU in Prague, Prague, Czech Republic
{kubalik,babuska}@ciirc.cvut.cz
[2] Department of Cybernetics, Faculty of Electrical Engineering,
CTU in Prague, Prague, Czech Republic
[3] Delft Center for Systems and Control,
Delft University of Technology, Delft, The Netherlands

Abstract. This paper presents a first step of our research on designing an effective and efficient GP-based method for symbolic regression. First, we propose three extensions of the standard Single Node GP, namely (1) a selection strategy for choosing nodes to be mutated based on depth and performance of the nodes, (2) operators for placing a compact version of the best-performing graph to the beginning and to the end of the population, respectively, and (3) a local search strategy with multiple mutations applied in each iteration. All the proposed modifications have been experimentally evaluated on five symbolic regression benchmarks and compared with standard GP and SNGP. The achieved results are promising showing the potential of the proposed modifications to improve the performance of the SNGP algorithm. We then propose two variants of hybrid SNGP utilizing a linear regression technique, LASSO, to improve its performance. The proposed algorithms have been compared to the state-of-the-art symbolic regression methods that also make use of the linear regression techniques on four real-world benchmarks. The results show the hybrid SNGP algorithms are at least competitive with or better than the compared methods.

Keywords: Genetic programming · Single node genetic programming · Symbolic regression

1 Introduction

This paper presents a first step of our research on genetic programming (GP) for the symbolic regression problem. The ultimate goal of our project is to design an effective and efficient GP-based method for solving dynamic symbolic regression problems where the target function evolves in time. Symbolic regression (SR) is a type of regression analysis that searches the space of mathematical expressions to find the model that best fits a given dataset, both in terms of accuracy and simplicity[1].

---

[1] https://en.wikipedia.org/wiki/Symbolic_regression.

© Springer-Verlag Berlin Heidelberg 2016
N.T. Nguyen et al. (Eds.): TCCI XXIV, LNCS 9770, pp. 61–82, 2016.
DOI: 10.1007/978-3-662-53525-7_4

Genetic programming belongs to effective and efficient methods for solving the SR problem. Besides the standard Koza's tree-based GP [12], many other variants have been proposed. They include, for instance, Grammatical Evolution (GE) [20] which evolves programs whose syntax is defined by a user-specified grammar (usually a grammar in Backus-Naur form). Gene Expression Programming (GEP) [4] is another GP variant successful in solving the SR problems. Similarly to GE it evolves linear chromosomes that are expressed as tree structures through a genotype-phenotype mapping. A graph-based Cartesian GP (CGP) [18], is a GP technique that uses a very simple integer based genetic representation of a program in the form of a directed graph. In its classic form, CGP uses a variant of a simple algorithm called $(1 + \lambda)$-Evolution Strategy with a point mutation variation operator. When searching the space of candidate solutions, CGP makes use of so called *neutral mutations*, meaning that a move to the new state is accepted if it does not worsen the quality of the current solution. This allows an introduction of new pieces of genetic code that can be plugged into the functional code later on and allows for traversing plateaus of the fitness landscape.

A Single Node GP (SNGP) [9,10] is a rather new graph-based GP system that evolves a population of individuals, each consisting of a single program node. Similarly to CGP, the evolution is carried out via a hill-climbing mechanism using a single reversible mutation operator. The first experiments with SNGP were very promising as they showed that SNGP significantly outperforms the standard GP on various problems including the SR problem. In this work we take the standard SNGP as the baseline approach and propose several modifications to further improve its performance.

The goals of this work are twofold. The first goal is to verify performance of the vanilla SNGP compared to the standard GP on various SR benchmarks and to investigate the impact of the following three design aspects of the SNGP algorithm:

- a strategy to select the nodes to be mutated,
- a strategy according to which the nodes of the best-performing expression are treated in the population,
- and a type of the search strategy used to guide the optimization process.

The second goal is to propose a hybrid variant of SNGP which incorporates the LASSO regression technique for creating linear-in-parameters nonlinear models. We compare its performance with other state-of-the-art symbolic regression methods which also make use of linear regression techniques.

The paper is organized as follows. Section 2 describes the SNGP algorithm. In Sect. 3, three modifications of the SNGP algorithm are proposed. Experimental evaluation of the modified SNGP and its comparison to the standard SNGP and standard Koza's GP is presented in Sect. 4. Section 5 describes two variants of the SNGP utilizing the linear regression technique, LASSO, to improve its performance. The two versions of SNGP with LASSO are compared to other symbolic regression methods making use of the linear regression techniques in Sect. 6. Finally, Sect. 7 concludes the paper and proposes directions for the further research on this topic.

2 Single Node Genetic Programming

2.1 Representation

The Single Node Genetic Programming is a GP system that evolves a population of individuals, each consisting of a single program node. The node can be either terminal, i.e. a constant or a variable node, or a function from a set of functions defined for the problem at hand. Importantly, individuals are not isolated in the population, they are interlinked in a graph structure similar to that of CGP, with population members acting as operands of other members [9].

Formally, a SNGP population is a set of N individuals $M = \{m_0, m_1, \ldots, m_{N-1}\}$, with each individual m_i being a single node represented by the tuple $m_i = \langle u_i, f_i, Succ_i, Pred_i, O_i \rangle$, where

- $u_i \in T \cup F$ is either an element chosen from a function set F or a terminal set T defined for the problem,
- f_i is the fitness of the individual,
- $Succ_i$ is a set of successors of this node, i.e. the nodes whose output serves as the input to the node,
- $Pred_i$ is a set of predecessors of this node, i.e. the nodes that use this individual as an operand,
- O_i is a vector of outputs produced by this node.

Typically, the population is partitioned so that the first N_{term} nodes, at positions 0 to $N_{term} - 1$, are terminals (variables and constants in case of the SR problem), followed by function nodes. Importantly, a function node at position i can use as its successor (i.e. the operand) any node that is positioned lower down in the population relative to the node i. This means that for each $s \in Succ_i$ we have $0 \leq s < i$ [9]. Similarly, predecessors of individual i must occupy higher positions in the population, i.e. for each $p \in Pred_i$ we have $i < p < N$. Note that each function node is in fact a root of a *direct acyclic graph* that can be constructed by recursively traversing through successors until the leaf terminal nodes.

2.2 Evolutionary Model

In [9], a single evolutionary operator called *successor mutate* (*smut*) has been proposed. It picks one individual of the population at random and then one of its successors is replaced by a reference to another individual of the population making sure that the constraint imposed on the successors is satisfied. Predecessor lists of all affected individuals are updated accordingly. Moreover, all individuals affected by this action must be reevaluated as well. For more details refer to [9].

The evolution is carried out via a hill-climbing mechanism using a *smut* operator and an acceptance rule, which can have various forms. In [9], it was based on fitness measurements across the whole population, rather than on single individuals. This means that once the population has been changed by a single application of the *smut* operator and all affected individuals have been re-evaluated, the new population is accepted if and only if the sum of the fitness values of all individuals in the population is no worse than the sum of fitness values before the mutation. Otherwise, the modifications made by the mutation are reversed. In [10] the acceptance rule is based only on the best fitness in the population. The latter acceptance rule will be used in this work as well. The reason for this choice is explained in Sect. 3.4.

3 Proposed Modifications

In this section, the following three modifications of the SNGP algorithm will be proposed:

1. A selection strategy for choosing nodes to be mutated based on depth and performance of nodes.
2. Operators for placing a compact version of the tree rooted in the best performing node to the beginning and to the end of the population, respectively.
3. A local search strategy with multiple mutations applied in each iteration.

In the following text, the term "best tree" is used to denote the tree rooted in the best performing node.

3.1 Depthwise Selection Strategy

The first modification focuses on the strategy for selecting the nodes to be mutated. In the standard SNGP, the node to be mutated is chosen at random. This means that all function nodes have the same probability of selection irrespectively of (1) how well they are performing and (2) how well the trees of which they are a part are performing. This is not in line with the evolutionary paradigm where the well fit individuals should have higher chance to take part in the process of an evolution of the population.

One way to narrow this situation is to select nodes according to their fitness. However, this would prefer just the root nodes of trees with high fitness while neglecting the nodes at the deeper levels of such well-performing trees which themselves have rather poor fitness. In fact, imposing high selection pressure on the root nodes might be counter-productive in the end as the mutations applied on the root nodes are less likely to bring an improvement than mutations applied on the deeper structures of the trees.

We propose a selection strategy that takes into account the quality of the mutated trees, so that better performing trees are preferred, as well as the depth of the mutated nodes so that deeper nodes of the trees are preferred to the shallow ones. The selection procedure has four steps:

1. A function node n is chosen at random.
2. A tree t with the best fitness out of all trees that use the node n is chosen.
3. All nodes of the tree t are collected in a set S. Each node is assigned a score equal to its depth in the tree t.
4. One node is chosen from the set S using a binary tournament selection considering the score values in the higher the better manner.

3.2   Organization of the Population

The second modification aims at improving the exploration capabilities of the SNGP algorithm. Two operators for placing a compact version of the best performing graph to the beginning and to the end of the population, respectively, are proposed.

Move left operator. Let us first describe the motivation for and the realization of the operator that places the compact version of the best graph to the beginning of the population. The motivation for this operator, denoted as *moveLeft* operator, is that well-performing nodes (and the whole graph structure rooted in this node) can represent a suitable building block for constructing even better trees when used as a successor of other nodes in the population. Since the chance of any node of being selected as a successor is higher if the node is more to the left in the population, it might be beneficial to store the well-performing graphs at lower positions in the population. Thus, the operator takes the best graph, G_{best}, and places it in a *compact form* to the very beginning of the population. By the compact form of a graph G we mean a sequence of nodes representing the whole G such that it contains only nodes involved in G. The *moveLeft* operator works as follows:

1. Extract nodes of the graph G_{best} rooted in the best-performing node and put the nodes into a compact ordered list L.
2. Set all successor and predecessor links of nodes within L so that L represents the same graph as the original graph G_{best}.
3. Place L to the beginning of the population, i.e. the first node of L being at the first function node position in the population.
4. Update the successor links of nodes of the original graph G_{best} so that it retains the same functionality as it had before the action.

    It must be made sure that all nodes of the original G_{best} have properly set their successors. If for example some successor of a node of the original G_{best} gets modified (i.e. the successor falls into the portion of the population newly occupied by the compact form of the G_{best}), then the successor reference is updated accordingly. In Fig. 1, this is for example the case of the second successor of the node at position 8, which originally pointed to the node number 4 and after the *moveLeft* operation has been redirected to the node number 2.
5. Update the predecessor lists of nodes in the compact form of G_{best}) in order to reestablish links to other nodes in the population that use the nodes as successors.

    In the example in Fig. 1 this is the case of the predecessor number 7 of node number 4 and the predecessor number 9 of node number 5, respectively.

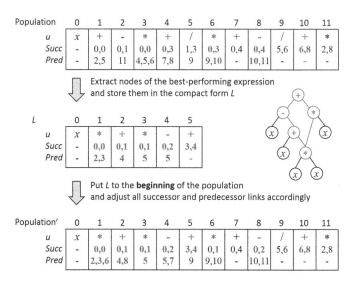

Population	0	1	2	3	4	5	6	7	8	9	10	11
u	x	+	-	*	+	/	*	+	-	/	+	*
Succ	-	0,0	0,1	0,0	0,3	1,3	0,3	0,4	0,4	5,6	6,8	2,8
Pred	-	2,5	11	4,5,6	7,8	9	9,10	-	10,11	-	-	-

Extract nodes of the best-performing expression and store them in the compact form L

L	0	1	2	3	4	5
u	x	*	+	*	-	+
Succ	-	0,0	0,1	0,1	0,2	3,4
Pred	-	2,3	4	5	5	-

Put L to the **beginning** of the population and adjust all successor and predecessor links accordingly

Population'	0	1	2	3	4	5	6	7	8	9	10	11
u	x	*	+	*	-	+	*	+	-	/	+	*
Succ	-	0,0	0,1	0,1	0,2	3,4	0,1	0,4	0,2	5,6	6,8	2,8
Pred	-	2,3,6	4,8	5	5,7	9	9,10	-	10,11	-	-	-

Fig. 1. Illustration of the *moveLeft* operator. Function nodes involved in the original and compact form of the best graph are shown in red. After the application of the *moveLeft* operator the population contains two occurrences of the best graph, the one represented by a sequence of nodes [0, 1, 2, 6, 8, 10] and the one represented by nodes [0, 1, 2, 3, 4, 5]. (Color figure online)

Note that after the application of the *moveLeft* operator the population contains two versions of the G_{best}, the original one and the compact one, see the example in Fig. 1.

Move right operator. Similarly, an operator that places the compact version of the best graph G_{best} to the end of the population is proposed. The motivation for this operator, denoted as *moveRight* operator, is that a performance of some well-performing graphs can more likely be improved by mutations applied to the nodes on deeper levels of the graph than by mutations applied to the root node or shallow nodes of the graph. In order to increase the number of possible structural changes to the deeper nodes of the best graph the compact version of the graph is placed to the end of the population. The working scenario for the operator is similar to the one of the *moveLeft* operator, see Fig. 2. Note that the application of the *moveRight* operator might result in the final population that contains just a single occurrence of the G_{best}. This might happen when the nodes of the original G_{best} fall into the area of its compact form.

3.3 Local Search Strategy

The last modification of the standard SNGP algorithm consists in allowing multiple mutation in a single iteration of the local search procedure. The idea behind this modification is rather straightforward. During the course of the optimization process the population might converge to the local optimum state where it is

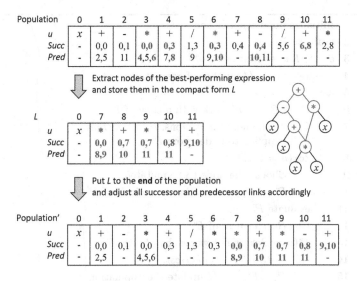

Population	0	1	2	3	4	5	6	7	8	9	10	11
u	x	+	-	*	+	/	*	+	-	/	+	*
Succ	-	0,0	0,1	0,0	0,3	1,3	0,3	0,4	0,4	5,6	6,8	2,8
Pred	-	2,5	11	4,5,6	7,8	9	9,10	-	10,11	-	-	-

Extract nodes of the best-performing expression and store them in the compact form L

L	0	7	8	9	10	11
u	x	*	+	*	-	+
Succ	-	0,0	0,7	0,7	0,8	9,10
Pred	-	8,9	10	11	11	-

Put L to the **end** of the population and adjust all successor and predecessor links accordingly

Population'	0	1	2	3	4	5	6	7	8	9	10	11
u	x	+	-	*	+	/	*	*	+	*	-	+
Succ	-	0,0	0,1	0,0	0,3	1,3	0,3	0,0	0,7	0,7	0,8	9,10
Pred	-	2,5	-	4,5,6	-	-	-	8,9	10	11	11	-

Fig. 2. Illustration of the *moveRight* operator. Function nodes involved in the original and compact version of the best graph are shown in red. In this example, the final population contains just one occurrence of the best graph represented by a sequence of nodes [0, 7, 8, 9, 10, 11]. (Color figure online)

hard to find further improvement by just one application of the smut operator. With multiple mutations applied in each iteration, the probability of getting stuck in such local optimal should be reduced. In this work, a parameter $upToN$ specifying the maximum number of mutation applications is used. Thus, if the parameter is set for example to 5, a randomly chosen number from interval $\langle 1, 5 \rangle$ of mutations are applied to the population in each iteration.

3.4   Outline of Modified SNGP Algorithm

This section presents an outline of the generic SNGP algorithm with possible utilization of the proposed modifications, see Fig. 3. In each generation, k mutations are applied to nodes of the population, see steps 8–10. In case of the standard SNGP just a single mutation is applied in each generation. After all k mutations have been applied, the nodes affected by this action gets reevaluated. If the best fitness of the modified population is not worse than the current best-so-far fitness than the modified population becomes the current population for the next generation, see step 15. Here, the fitness of each individual is calculated as the **sum of absolute errors** (SAE) generated by the individual over all training samples. In step 16, the operators moving the best tree to the beginning or to the end of the population are applied to the population, if applicable. Then the fitness evaluation counter is incremented and if there are still some fitness evaluations left the next generation is carried out. Once the maximal number of fitness evaluations is used the best node (and its tree) of the population is returned.

```
 1  Initialize population of nodes, P
 2  evaluate P
 3  bestSoFar ← best of P
 4  i ← 0              // number of generations
 5  do
 6      P' ← P        // work with a copy of P
 7      Choose the no. of mutations, k ∈ (1, upToN),
        to be applied to nodes from P'
 8      for(n = 1 ... k)
 9          Choose the node to be mutated, N
10          P' ← Apply mutation to node N of P'
11      evaluate P'
12      currBest ← best of P'
13      if(currBest is not worse than bestSoFar)
14          bestSoFar ← currBest
15          P ← P'     // update the population
16          If applicable, apply either moveLeft
            or moveRight operator to P
17      i ← i + 1
18  while (i ≤ maxGeneration)
19  return bestSoFar
```

Fig. 3. Outline of the modified SNGP algorithm.

In this work, we use the acceptance criterion, step 13, working with the best fitness in the population, not the average fitness of the population. The reason is that when the *moveLeft* and *moveRight* operators are used, they might significantly change the average fitness of the population while the best fitness stays intact.

4   Experiments with Modified SNGP

This section presents experiments carried out with standard GP, standard SNGP and SNGP with the proposed modifications.

4.1   Artificial Benchmarks

The algorithms have been tested on five symbolic regression benchmarks

- $f_1(x) = 4x^4 - 3x^3 + 2x^2 - x$,
  32 training samples equidistantly sampled from $\langle 0, 1.0 \rangle$,
- $f_2(x) = x^6 - 2x^4 + x^2$,
  100 training samples equidistantly sampled from $\langle -1.0, 1.0 \rangle$

- $f_3(x) = x^6 - 2.6x^4 + 1.7x^2$,
  100 training samples equidistantly sampled from $\langle -1.0, 1.0 \rangle$
- $f_4(x) = x^6 - 2.6x^4 + 1.7x^2 - x$,
  100 training samples equidistantly sampled from $\langle -1.4, 1.4 \rangle$
- $f_5(x_1, x_2) = \frac{(x_1-3)^4 + (x_2-3)^3 - (x_2-3)}{(x_2-2)^4 + 10}$,
  100 training samples equidistantly sampled from $\langle 0.05, 6.05 \rangle$

The first two functions are rather simple polynomials with small integer constants. We chose the function f_1 since it was used in the original SNGP paper [9]. Function f_2 is the Koza-3 function taken from [17]. Functions f_3 and f_4 are modifications of the Koza-3 function so that they involve non-trivial decimal constants. Thus, these functions should represent harder instances than f_1 and f_2. The function f_4 is made even harder than f_3 while breaking the symmetry by adding the term "$-x$". The last function f_5 is a representative of a rational function of two variables. This function, known as Vladislavleva-8 function [17], represents the hardest SR problem used in this work.

4.2 Experimental Setup

All the tested variants of the SNGP use a population of size 400. The population starts with terminal nodes representing the variable x_1 and x_2 and a constant 1.0 followed by function nodes of types $\{+, -, *, /\}$. SNGP was run for 25,000 iterations, in each iteration just a single population reevaluation is computed (note, just the nodes that were affected by the mutation are reevaluated). The number of iterations was chosen so as to make the comparisons of the GP and SNGP as fair as possible. This way a balance between processed nodes and fitness evaluations is found, see [19].

The proposed modifications of the SNGP algorithm are configured with the following parameters:

- $upToN \in \{1, 5\}$,
- *selection* is either random (denoted as 'r') or depthwise (denoted as 'd')
- *moveType* is either *moveLeft* (denoted as 'l'), *moveRight* (denoted as 'r') or no move (denoted as 'n').

Names of the tested configurations of the SNGP are constructed as follows "SNGP_*upToN_selection_moveType*". The standard SNGP is denoted as SNGP_1_r_n, i.e. SNGP with a random selection and no move operator applying a single mutation per generation.

Standard GP with generational replacement strategy was used with the following parameters:

- Function set: $\{+, -, *, /\}$
- Terminal set: $\{x_1, x_2, 1.0\}$
- Population size: 500
- Initialization method: Ramped half-and-half
- Tournament selection: 5 candidates

– Number of generations: 55, i.e. 54 generations plus initialization of the whole population
– Crossover probability: 90 %
– Reproduction probability: 10 %
– Probability of choosing internal node as crossover point: 90 %

For the experiments with the GP we used the Java-based Evolutionary Computation Research System ECJ 22[2].

One hundred independent runs were carried out with each tested algorithm on each benchmark and the observed performance characteristics are

– *fitness* – the mean best fitness (i.e. the sum of absolute errors) over 100 runs;
– *sample rate* – the mean number of successfully solved samples by the best-fitted individual calculated over 100 runs, where the sample is considered to be successfully solved by the individual iff the absolute error achieved by the individual on this sample is less then 0.01;
– *solution rate* – the percentage of complete solutions found within 100 runs, where the runs completely solves the problem iff the best individual generates on all training samples the absolute error less than 0.01;
– *size* – the mean number of nodes of the best solution found calculated over 100 runs.

4.3   Results

Results obtained with the compared algorithms are presented in Table 1. The first observation is that the results obtained on the benchmark f_1 are quite different than the results presented in [9], as the performance of the SNGP is not as good as the SNGP performance presented there whilst the standard GP performs much better than presented in [9]. This might be caused by different configurations of the SNGP and GP used in our work and in [9]. We used different acceptance criterion in SNGP and the generational instead of the steady-state replacement strategy in GP. This observation might indicate that both approaches are quite sensitive to the proper setting of their individual components.

The second observation is that the modified versions of SNGP systematically outperform the standard SNGP with respect to the *fitness, sample rate* and *solution rate* performance measures. On the other hand, the modified SNGP is not a clear winner over the standard GP. The SNGP outperforms GP on f_2, f_4 and f_5. On f_1 it performs equally well as the GP. On f_3, all versions of SNGP get outperformed by the GP with respect to the *fitness*. It turns out functions f_3, f_4 and f_5 represent a real challenge for all tested algorithms since no one was able to find a single correct solution within the 100 runs. We hypothesize the hardness of f_3 and f_4 stems from the fact these benchmarks involve non-trivial constants that might be hard to evolve. Function f_5 is hard since it is a rational function.

---

[2] https://cs.gmu.edu/~eclab/projects/ecj/.

Table 1. Results of the modified SNGP variants and standard GP on artificial benchmarks $f_1 - f_5$. The best mean fitness value for each benchmark is highlighted.

Function	Algorithm	Fitness	Sample rate (%)	Solution rate (%)	Nodes
f_1	GP	**0.14**	82.8	49	151
	SNGP_1_r_n	0.65	37.2	5	26.8
	SNGP_1_d_n	0.29	63.4	18	33.7
	SNGP_1_r_l	0.62	38.1	3	22.5
	SNGP_1_d_l	<u>0.25</u>	68.4	24	33.9
	SNGP_1_r_r	0.66	35.9	4	34.5
	SNGP_1_d_r	0.28	66.9	17	56.6
	SNGP_5_d_n	0.16	78.8	49	32.3
	SNGP_5_d_l	0.17	77.5	49	28.5
	SNGP_5_d_r	**0.14**	84.7	53	52.4
f_2	GP	0.78	88.7	69	175.1
	SNGP_1_r_n	0.85	73.4	33	27.7
	SNGP_1_d_n	0.17	94.4	78	27.6
	SNGP_1_r_l	0.75	78	33	30.8
	SNGP_1_d_l	0.25	92.8	65	27.7
	SNGP_1_r_r	0.84	72.7	17	47
	SNGP_1_d_r	0.15	94.3	82	40.9
	SNGP_5_d_n	**1e-6***	100	100	22.6
	SNGP_5_d_l	0.08	97.8	87	21.2
	SNGP_5_d_r	0.05	98.2	91	37.2
f_3	GP	**1.19**	68.4	0	155
	SNGP_1_r_n	2.7	40	0	28.9
	SNGP_1_d_n	1.5	54.0	0	33.6
	SNGP_1_r_l	<u>2.39</u>	43.7	0	30.1
	SNGP_1_d_l	1.4	59.2	0	35
	SNGP_1_r_r	2.75	37.2	0	43.4
	SNGP_1_d_r	1.6	51.6	0	56.2
	SNGP_5_d_n	1.37	59.2	0	32.6
	SNGP_5_d_l	<u>1.23</u>	65.6	0	30.3
	SNGP_5_d_r	1.35	60.8	0	57.2
f_4	GP	11.0	19.4	0	146.8
	SNGP_1_r_n	10.5	12.9	0	26.1
	SNGP_1_d_n	7.8	21.7	0	34.2
	SNGP_1_r_l	11.2	10.8	0	26.5
	SNGP_1_d_l	8.3	19.0	0	32.9
	SNGP_1_r_r	<u>8.9</u>	19.0	0	45.8
	SNGP_1_d_r	7.4	22.4	0	53.9
	SNGP_5_d_n	7.15	25.8	0	31.5
	SNGP_5_d_l	7.4	24.0	0	28
	SNGP_5_d_r	**6.7**	27.2	0	50.8
f_5	GP	71.2	4.4	0	194.3
	SNGP_1_r_n	64.1	4.0	0	26.0
	SNGP_1_d_n	61.6	3.8	0	35.6
	SNGP_1_r_l	64.9	3.8	0	27.3
	SNGP_1_d_l	**60.0**	4.1	0	34.9
	SNGP_1_r_r	63.6	4.4	0	44.3
	SNGP_1_d_r	61.1	4.1	0	51.6
	SNGP_5_d_n	60.7	3.7	0	32.8
	SNGP_5_d_l	60.9	3.6	0	31.9
	SNGP_5_d_r	60.4	4.0	0	51.1

The third observation is that there is a clear trend showing that the depthwise node selection works significantly better than the random one. Whenever the SNGP configurations differ just in the selection type the one using the depthwise selection outperforms the one with the random selection.

The fourth observation is that the reorganization of the population using either the *moveLeft* or *moveRight* operator does not have any systematic impact on the overall performance of the algorithm. It happens only rarely that the SNGP using *moveLeft* or *moveRight* outperforms its counterpart configuration with no move operator used. In particular, the *moveLeft* operator was significantly better[3] than *noMove* in four cases, the *moveRight* operator was significantly better than *noMove* in two cases, all the cases indicated by underlined values. On the other hand, the *noMove* configuration happened to outperform both the *moveLeft* and *moveRight* configuration on function f_2 as indicated by an asterisk.

The fifth observation is that the local search strategy allowing multiple mutations in one iteration outperforms the standard local search procedure with just a single application of the mutation operator per iteration. This is with agreement with our expectations.

Last but not least, the SNGP consistently finds much smaller trees than the GP. This is very important since very often solutions of small size that are interpretable by human are sought in practice.

5   Hybrid SNGP with Linear Regression

It has widely been reported in the literature that the evolutionary algorithms work much better when hybridized with local search techniques, the concept known as the memetic algorithms [7]. EA serves as a global search strategy, while the local search technique provides an efficient means for fine-tuning the solutions. A similar approach can be used to develop efficient methods for symbolic regression.

Recently, several methods emerged [1, 2, 15, 21, 22] that explicitly restrict the class of models to generalized linear models, i.e. to a linear combination of possibly non-linear basis functions. With the help of linear regression techniques applied to the basis functions, such models can be learned much faster.

GPTIPS [21, 22] is an open-source SR toolbox for MATLAB. It is an implementation of Multi-Gene Genetic Programming (MGGP) [8] and thus has its roots in classical GP. Each solution is composed of multiple independent trees, called genes, and their outputs are linearly combined. The coefficients of this linear combination are computed optimally with respect to the MSE of the final output to the true target values by classical least-squares linear regression. GPTIPS (MGGP) is based on classical Genetic Program- ming. This means that it works with a population of fixed size, subtree mutation, subtree crossover, tournament selection, standard initialization procedures, and is able to handle

---

[3] Checked using the t-test calculated with the significance level $\alpha = 0.05$.

the internal constants of the model (to certain extent) using ephemeral random constants. The output of GPTIPS is a population of models; it is up to the user to choose the final one. By default, GPTIPS uses Lexicographic Parsimony Pressure [13] using (by default) Expressional Complexity [24] of the models. MGGP was shown to be faster and more accurate than conventional GP [8] and also a comparable or better alternative to classical methods like Support Vector Regression and Artificial Neural Networks [6].

FFX, or Fast Function Extraction [2], is a deterministic algorithm for symbolic regression. It first exhaustively generates a massive set of basis functions, which are then linearly combined using Pathwise Regularized Learning [5,25] to produce sparse models. The algorithm produces a Pareto-front of models with respect to their accuracy and complexity. Again, it is up to the user to choose the final model. There are two kinds of bases that are generated: univariate bases and bivariate bases. Univariate bases are: a variable raised to a power (chosen from a fixed set of options) and (non-linear) functions applied to another univariate base. Bivariate bases are products of all pairs of univariate bases excluding the pairs where both the bases are of function-type; the author argues that such products are "deemed to be complex". FFX also includes a trick that allows it to produce rational functions of the bases using the same learning procedure. The original paper reports FFX to be more accurate than many classical methods including conventional GP, neural networks and SVM.

EFS, or Evolutionary Feature Synthesis [2] is a recent evolutionary-based algorithm. In EFS, the population does not consist of complete models but rather of features which, collectively, form a single model. The initial population is formed by the original features of the dataset. Then, in each generation, a model is composed of the features in the current population by Pathwise Regularized Learning and is stored if it is the best. The next step in a generation is the composition of new features by applying unary and binary functions to the features already present in the current population. This way, more complex features are created from simpler ones. Also, the features are selected during this composition step according to the Pearson correlation coefficient with the feature's parents. EFS does not build the symbolic model explicitly – it works with the data of the features in a vectorial fashion and only stores the structure for logging purposes. This results in a very fast algorithm. The original paper reports EFS being comparable to neural networks and similar or better than Multiple Regression Genetic Programming [1] which itself was reported to outperform conventional GP, multiple regression and Scaled Symbolic Regression [11].

In this section we propose two variants of hybrid SNGP that make use of the linear regression technique to improve its performance. Both use the Least Absolute Shrinkage and Selection (LASSO) regression technique, the one used in EFS, to build generalized linear regression models. The first one, denoted as *Single-Run SNGP with LASSO* (s-SNGPL), evolves a population of candidate features for the LASSO regression in a single run of the SNGP. The second variant, denoted as *Iterated SNGP with LASSO* (i-SNGPL), builds the LASSO model in an iterative manner where in each iteration a new feature for the LASSO model is evolved in a separate SNGP run.

5.1   Single-Run SNGP with LASSO

In this method, all features of the generalized linear regression model are evolved in a single run of the SNGP. The outline of the algorithm, see Fig. 4, is very much like the one of the modified SNGP, see Fig. 3. The only difference is in the evaluation of individual nodes in the population and in assessment of the overall population's quality after the content of the population has been altered in each generation. First, a quality of each node is calculated as the Pearson product-moment correlation coefficient between the node's output and the desired output values (line 11). Then, a generalized linear regression model of a subset of features present in the population is calculated using the LASSO technique (line 12). Finally, the fitness of the whole population is calculated as the **sum of absolute errors** between the LASSO regression model output and the desired output values. Thus, the hybrid SNGP uses the same fitness as the modified SNGP, see Sect. 3.4.

The complexity of the LASSO model is controlled by (1) the maximal depth of features evolved in the population and (2) the maximum number of features the LASSO model can be composed of. Note, the features can be non-linear functions.

5.2   Iterated SNGP with LASSO

Unlike the s-SNGPL, here the set of candidate features $\mathbf{F}$ for the LASSO regression model is not evolved within a single population of SNGP. Instead, an external set $\mathbf{F}$ is build incrementally, starting from an empty set and adding one feature in each iteration, see Fig. 5.

Each feature f_i is evolved in a separate run of the SNGP (line 6) such that it correlates the most with the residua $\mathbf{R}$ (i.e. the vector of error values over all training samples) produced by the current LASSO regression model composed of $i-1$ features. The residua are initialized to desired output values of the training samples. The idea is that in each iteration a new feature is evolved such that it possibly helps to reduce the error of the resulting LASSO model. The algorithm stops when either the set of candidate features reached the preset maximum or the error of the LASSO model becomes zero.

6   Experiments with Hybrid SNGP

First experiments with hybrid SNGP variants s-SNGPL and i-SNGPL are carried out on the artificial benchmarks listed in Sect. 4.1. Another series of experiments are carried out on real-world benchmarks described in the following section.

```
1    initialize population of nodes P
2    calculate fitness of nodes in P
     based on their correlation with expected outputs
3    build LASSO regression model LM(P)
4    i ← 0          // number of generations
5    do
6          P' ← P      // work with a copy of P
7          Choose the no. of mutations, k ∈ (1, upToN),
           to be applied to nodes from P'
8          for(n = 1 . . . k)
9                choose the node to be mutated, N
10               P' ← Apply mutation to node N of P'
11         calculate fitness of nodes in P'
           based on their correlation with expected outputs
12         build LASSO regression model LM(P')
13         if(LM(P') is not worse than LM(P))
14               P ← P'     // update the population
15         i ← i + 1
16   while (i ≤ maxGenerations)
17   return LM(P)
```

Fig. 4. Outline of the Single-Run SNGP with LASSO.

```
1    i ← 0       // number of candidate features
2    initialize feature set F = ∅
3    initialize residua R
4    do
5          i ← i + 1
6          evolve a new feature f_i using separate SNGP
           based on its correlation with R
7          add f_i to F
8          build Lasso model LM(F) using all features in F
9          update residua R
10   while (i ≤ maxFeatures and R ≠ 0)
11   return LM
```

Fig. 5. Outline of the Iterated SNGP with LASSO.

6.1   Real-World Benchmarks

Following four real-world benchmarks, acquired from the UCI repository [14], were used in this work

- **Energy Efficiency of Cooling (ENC) and Heating (ENH)** are datasets regarding the energy efficiency of cooling and heating of buildings. Dimension is 8, number of datapoints is 768.
- **Concrete Compressive Strength (CCS)** is a dataset representing a highly non-linear function of concrete age and ingredients. Dimension of the dataset is 8, the number of datapoints is 1030.
- **Airfoil Self-Noise (ASN)** is a dataset regarding the sound pressure levels of airfoils based on measurements from a wind tunnel. Dimension of the dataset is 5, the number of datapoints is 1503.

These benchmarks were used in the work on EFS [2] and other relevant literature.

Each dataset was split 100 times (using the 0.7/0.3 ratio for training/testing). Each algorithm was run once on each of the dataset instances producing a single model. The accuracy and complexity of the resulting models are then aggregated and statistically compared.

6.2   Experimental Setup

We compare the proposed hybrid SNGP algorithms to the GPTIPS, EFS and FFX. We used GPTIPS version 2 retrieved from [23], FFX in version 1.3.4 retrieved from [16], EFS was retrieved from [3]. The goal is to perform a comparison of the chosen methods as ready-to-use tools. Therefore we didn't modify to the code of the algorithms[4], and we left all of the settings at their default values. We set a timeout to 10 min for both EFS and GPTIPS. FFX has no support for timeout. However, the algorithms's performances have not been analyzed from the computation time point of view. No parameter tuning method was used to find an optimal configuration of the compared algorithms for particular benchmarks.

The most important for our evaluation purposes is how the algorithms control the resulting model complexity. GPTIPS has (user-defined) limits on the maximum number of nodes and/or maximum depth, and on the maximum number of bases. By default there is a depth limit of 4, and maximum number of bases (not counting the intercept) is also 4. EFS computes the maximum number of bases from the number of input features, p; the number of bases was set to $3p$ and maximum number of nodes in a base is hard-coded to 5. The FFX procedure results in a maximum model depth of 5.

The hybrid SNGP algorithms with LASSO regression were run on artificial benchmarks with the same population size and the sets of terminals and functions

---

[4] The only exception is EFS: we changed the round variable to false (which was originally hard-coded to true) according to the issue on the algorithm's GitHub repository, see https://github.com/exgp/efs/issues/1.

as were used in Sect. 4.2. The maximum number of generations of s-SNGPL was set to 1000. The maximum number of generations of each individual SNGP run of the i-SNGPL was set to 1000 as well. The maximum number of features the LASSO model can be composed of was set to 15 and the maximum depth of the evolved features was set to 4.

The modified SNGP and hybrid SNGP algorithms with LASSO regression were run on real-world benchmarks with the following changes in the configuration. The set of terminals was extended with constants 2.0, 3.0 and 4.0 and the set of functions was extended with functions square, cube, sqrt and sin. Similarly to EFS, the maximum number of features was set to $3p$, unless stated otherwise.

On the real-world benchmarks, we compare the resulting models with respect to the root mean square error (RMSE) and the number of nodes used in the model. We define the number of nodes as a sum of the numbers of nodes in the model's bases, i.e. we count neither the coefficients (including the intercept) of the linear combination, nor the multiplications between these coefficients and the bases. FFX's hinge functions, having a form $max(0; x - threshold)$ or similar, count as 5 nodes.

6.3   Results on Artificial Benchmarks

Table 2 shows results of the modified SNGP algorithm and the two hybrid SNGP algorithms using LASSO regression on the artificial benchmarks. Only the best performing configuration of the modified SNGP is selected for each benchmark based on the results presented in Table 1.

There is no single winner algorithm consistently outperforming the others on all five benchmarks. However, there is a clear trend showing that the SNGP without LASSO is doing well on rather simple benchmarks f_1 and f_2 (it is even better than both hybrid algorithms on f_2), i.e. the polynomials that involve only trivial integer constants. As the difficulty of the target model increases (from f_3 to f_5) the hybrid SNGP algorithms start to dominate. Of the two variants the i-SNGPL is better with respect to the SAE performance measure. Note, the superiority of the i-SNGPL is achieved at the cost of rather highly complex models, approximately 150 nodes and more compared to 65 to 118 nodes in case of s-SNGPL and 30 to 50 nodes in case of simple SNGP. These observations are in accordance with our expectations.

6.4   Results on Real-World Benchmarks

This section presents comparisons of the proposed modified and hybrid SNGP algorithms with GPTIPS, EFS, and FFX on the real-world benchmarks. The first observation based on results in Table 3 is that the simple SNGP without LASSO regression gets defeated by the other algorithms on all benchmarks.

The i-SNGPL outperforms the other algorithms on all benchmarks but the CCS, where the FFX exhibits the best median RMSE value. However, this is at the cost of very large models produced, see Table 4. Also the superiority of i-SNGPL on the three benchmarks is thank to large models produced by the

Table 2. Comparisons of the modified SNGP with two variants of hybrid SNGP using LASSO regression on artificial benchmarks $f_1 - f_5$. The best mean SAE value in each row is highlighted. For f_3, f_4 and f_5 the highlighted mean value was significantly better than the other two values as supported by the t-test calculated with the significance level $\alpha = 0.05$.

Function	Algorithm	SAE	Sample rate (%)	Solution rate (%)	Nodes
f_1	SNGP_5_d_r	0.14	84.7	53	52.4
	s-SNGPL	0.58	44.1	0	42.0
	i-SNGPL	**0.11**	97.5	54	74.8
f_2	SNGP_5_d_n	**1e-6**	100	100	22.6
	s-SNGPL	0.07	99.9	97	85.5
	i-SNGPL	0.34	93.6	44	124.5
f_3	SNGP_5_d_l	1.23	65.6	0	30.3
	s-SNGPL	**0.13**	99.5	84	83.6
	i-SNGPL	0.4	90.8	40	146.6
f_4	SNGP_5_d_r	6.7	27.2	0	50.8
	s-SNGPL	6.3	18.0	0	64.8
	i-SNGPL	**2.53**	34.5	0	147
f_5	SNGP_1_d_l	60.0	4.1	0	34.9
	s-SNGPL	27.8	3.3	0	117.9
	i-SNGPL	**15.9**	4.0	0	170.1

Table 3. Comparisons of SNGP_5_d_n, s-SNGPL and i-SNGPL with GPTIPS, EFS and FFX on the real-world benchmarks with respect to the median RMSE observed on testing data. The best value in each row is highlighted. In all cases the highlighted value was significantly better than the other values as supported by the Mann-Whitney U-test calculated with the significance level $\alpha = 0.01$.

	GPTIPS	EFS	FFX	SNGP_5_d_n	s-SNGPL	i-SNGPL
ENC	2.9073	1.6398	1.7906	3.5657	1.7076	**1.3978**
ENH	2.5375	0.5455	1.0455	3.4295	0.6583	**0.4754**
CCS	8.7618	6.4293	**5.9860**	10.55	6.4052	6.2144
ASN	4.1384	3.6232	3.5804	6.6852	6.1353	**2.9561**

algorithm. The s-SNGPL is competitive to the three compared algorithms with respect to the median RMSE as well as the model size.

Tables 5 and 6 show the performance of s-SNGPL and i-SNGPL achieved with smaller LASSO models. Values $k_1 \ldots k_4$ specify the maximum number of features the algorithms are allowed to use in the LASSO regression models. An interesting observation is that both algorithms, and especially the i-SNGPL one, stay competitive with the compared algorithms even when producing smaller models.

Table 4. Median number of nodes for each algorithm and dataset

	GPTIPS	EFS	FFX	SNGP_5_d_n	s-SNGPL	i-SNGPL
ENC	48	108	136	22	115.5	201
ENH	47.5	105	146	20.5	107	196.5
CCS	43	108	474.5	23	127	201
ASN	58	67	52.5	22.5	88	131

Table 5. Median RMSE and median number of nodes observed for **s-SNGPL** on testing data. Performance of the algorithms is tested for different values of the maximum number of features allowed for the LASSO model. Values $k_1 = 12$, $k_2 = 16$, $k_3 = 20$ and $k_4 = 24$ are tested on benchmarks ENC, ENH and CCS. Values $k_1 = 8$, $k_2 = 10$, $k_3 = 12$ and $k_4 = 15$ are tested on benchmark ASN.

	k_1		k_2		k_3		k_4	
	RMSE	#nodes	RMSE	#nodes	RMSE	#nodes	RMSE	nodes
ENC	1.8582	58.5	1.7694	82	1.6897	94	1.7076	115.5
ENH	1.0842	60	0.8461	75	0.8123	95	0.6583	107
CCS	6.8929	63	6.6912	82	6.5595	102	6.4053	127
ASN	4.0013	48	3.8395	61	3.7463	70	3.4817	88

Table 6. Median RMSE and median number of nodes observed for **i-SNGPL** on testing data. Performance of the algorithms is tested for different values of the maximum number of features allowed for the LASSO model. Values $k_1 = 12$, $k_2 = 16$, $k_3 = 20$ and $k_4 = 24$ are tested on benchmarks ENC, ENH and CCS. Values $k_1 = 8$, $k_2 = 10$, $k_3 = 12$ and $k_4 = 15$ are tested on benchmark ASN.

	k_1		k_2		k_3		k_4	
	RMSE	#nodes	RMSE	#nodes	RMSE	#nodes	RMSE	nodes
ENC	1.5490	101.5	1.4502	133	1.4085	170	1.3978	201.5
ENH	0.5648	97	0.5179	130.5	0.4978	166	0.4754	196.5
CCS	6.5912	103	6.4412	135	6.2973	167	6.2144	201
ASN	3.3894	71	3.2492	89.5	3.0989	106	2.9561	131

Figure 6 presents progress plots observed for the s-SNGPL algorithm on real-world benchmarks. In each generation, the mean of the best-so-far fitness (i.e. SAE) calculated over 100 independent runs is shown. It illustrates the effect of the evolutionary component of the algorithm as there is a clear continuous improvement in the best-so-far fitness along the whole run.

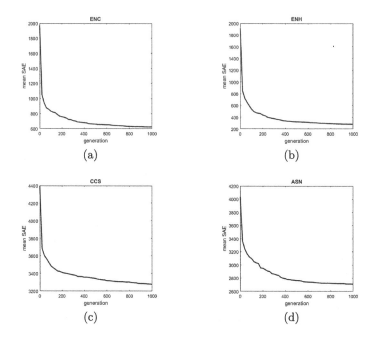

Fig. 6. Plots showing an average progress of the best SAE value for s-SNGPL on real-world benchmarks.

7   Conclusions

This paper deals with the Single Node Genetic Programming method, proposes its modifications and ways of hybridization to improve its performance.

First, three extensions of the standard SNGP, namely (1) a selection strategy for choosing nodes to be mutated based on the depth and performance of nodes, (2) operators for placing a compact version of the best-performing graph to the beginning and to the end of the population, respectively, and (3) a local search strategy with multiple mutations applied in each iteration were proposed.

These modifications have been experimentally evaluated on five artificial symbolic regression benchmarks and compared with standard GP and SNGP. The achieved results are promising showing the potential of the proposed modifications to improve the performance of the SNGP algorithm.

Further, two variants of hybrid SNGP utilizing the linear regression technique, LASSO, were proposed. The proposed hybrid algorithms have been compared to the state-of-the-art symbolic regression methods making use of the linear regression techniques on four real-world benchmarks. The results show the proposed algorithms are at least competitive with or better than the compared methods.

The next step of our research will be to carry out a thorough experimental evaluation of the modified SNGP algorithms with the primary objectives being the speed of convergence and the ability to react fast to the changes of the environment in order to be able to deploy the algorithm within the dynamic symbolic regression scenario. Further investigations will include utilization of new mutation operators, identification of suitable "high-level" basic functions to the SNGP's function set, design of mechanisms to evolve inner constants of the models and mechanisms for escaping from local optima.

Acknowledgment. This research was supported by the Grant Agency of the Czech Republic (GAČR) with the grant no. 15-22731S entitled "Symbolic Regression for Reinforcement Learning in Continuous Spaces".

References

1. Arnaldo, I., Krawiec, K., O'Reilly, U.-M.: Multiple regression genetic programming. In: Proceedings of the 2014 Annual Conference on Genetic and Evolutionary Computation, GECCO 2014, pp. 879–886. ACM, New York (2014)
2. Arnaldo, I., O'Reilly, U.-M., Veeramachaneni, K.: Building predictive models via feature synthesis. In: Proceedings of the 2015 Annual Conference on Genetic and Evolutionary Computation, GECCO 2015, pp. 983–990. ACM, New York (2015)
3. EFS commit 6d991fa. http://github.com/exgp/efs/tree/6d991fa
4. Ferreira, C.: Gene expression programming: a new adaptive algorithm for solving problems. Complex Syst. **13**(2), 87–129 (2001)
5. Friedman, J., Hastie, T., Tibshirani, R.: Regularization paths for generalized linear models via coordinate descent. J. Stat. Softw. **33**(1), 1–22 (2010)
6. Garg, A., Garg, A., Tai, K.: A multi-gene genetic programming model for estimating stress-dependent soil water retention curves. Comput. Geosci. **18**(1), 45–56 (2013)
7. Hart, E., Smith, J.E., Krasnogor, N.: Recent Advances in Memetic Algorithms. STUDFUZZ, vol. 166. Springer, Heidelberg (2005)
8. Hinchliffe, M., Hiden, H., McKay, B., Willis, M., Tham, M., Barton, G. Modelling chemical process systems using a multi-gene genetic programming algorithm. In: Koza, J.R. (ed.) Late Breaking Papers at the Genetic Programming 1996 Conference, pp. 56–65 (1996)
9. Jackson, D.: A new, node-focused model for genetic programming. In: Moraglio, A., Silva, S., Krawiec, K., Machado, P., Cotta, C. (eds.) EuroGP 2012. LNCS, vol. 7244, pp. 49–60. Springer, Heidelberg (2012). doi:10.1007/978-3-642-29139-5_5
10. Jackson, D.: Single node genetic programming on problems with side effects. In: Coello, C.A.C., Cutello, V., Deb, K., Forrest, S., Nicosia, G., Pavone, M. (eds.) PPSN 2012. LNCS, vol. 7491, pp. 327–336. Springer, Heidelberg (2012). doi:10.1007/978-3-642-32937-1_33
11. Keijzer, M.: Scaled symbolic regression. Genet. Program Evolvable Mach. **5**(3), 259–269 (2004)
12. Koza, J.: On the Programming of Computers by Means of Natural Selection, 2nd edn. MIT Press, Cambridge (1992)
13. Luke, S., Panait, L.: Lexicographic parsimony pressure. In: Proceedings of GECCO 2002, pp. 829–836. Morgan Kaufmann Publishers (2002)
14. Lichman, M.: UCI Machine Learning Repository. University of California, School of Information and Computer Science, Irvine (2013). http://archive.ics.uci.edu/ml
15. McConaghy, T.: Fast, scalable, deterministic symbolic regression technology. In: Riolo, R., Vladislavleva, E., Moore, J.H. (eds.) Genetic Programming Theory and Practice IX, Genetic and Evolutionary Computation, pp. 235–260 (2011)

16. FFX 1.3.4. http://pypi.python.org/pypi/ffx/1.3.4
17. McDermott, J., et al.: Genetic programming needs better benchmarks. In: Proceedings of the GECCO 2012, pp. 791–798. ACM, New York (2012)
18. Miller, J.F., Thomson, P.: Cartesian genetic programming. In: Poli, R., Banzhaf, W., Langdon, W.B., Miller, J., Nordin, P., Fogarty, T.C. (eds.) EuroGP 2000. LNCS, vol. 1802, pp. 121–132. Springer, Heidelberg (2000). doi:10.1007/978-3-540-46239-2_9
19. Ryan, C., Azad, R.M.A.: A simple approach to lifetime learning in genetic programming-based symbolic regression. Evol. Comput. **22**(2), 287–317 (2014)
20. Ryan, C., Collins, J.J., Neill, M.O.: Grammatical evolution: evolving programs for an arbitrary language. In: Banzhaf, W., Poli, R., Schoenauer, M., Fogarty, T.C. (eds.) EuroGP 1998. LNCS, vol. 1391, pp. 83–96. Springer, Heidelberg (1998). doi:10.1007/BFb0055930
21. Searson, D.P., Leahy, D.E., Willis, M.J.: Gptips: an open source genetic programming toolbox for multigene symbolic regression. In International MultiConference of Engineers and Computer Scientists, vol. 1, pp. 77–80 (2010)
22. Searson, D.P.: GPTIPS 2: an open-source software platform for symbolic datamining. In: Gandomi, A.H., Alavi, A.H., Ryan, C. (eds.) Springer Handbook of Genetic Programming Applications, pp. 551–573. Springer, Switzerland (2015)
23. GPTIPS 2. http://sites.google.com/site/gptips4matlab
24. Vladislavleva, E.J., Smits, G.F., Den Hertog, D.: Order of nonlinearity as a complexity measure for models generated by symbolic regression via pareto genetic programming. Trans. Evol. Comp. **13**(2), 333–349 (2009)
25. Zou, H., Hastie, T.: Regularization and variable selection via the elastic net. J. Roy. Stat. Soc. Ser. B (Stat. Methodol.) **67**(2), 301–320 (2005)

L2 Designer

A Tool for Genetic L-system Programming in Context of Generative Art

Tomáš Konrády[⊠], Kamila Štekerová, and Barbora Tesařová

Faculty of Informatics and Management, University of Hradec Králové,
Rokitanského 62, Hradec Králové, Czech Republic
{tomas.konrady,kamila.stekerova,
barbora.tesarova}@uhk.cz

Abstract. We propose a new format to define parametric L-systems (*L2 language*) and its implementation in JavaScript (*L2 Designer*). Our language allows us to create formal definition of the hierarchy of L-systems. The L2 Designer enables us to discover L-system grammars by means of interactive evolution - the common method used in Evolutionary art.

We provide an example of L2 program and we illustrate possibilities of L2 Designer on the two case studies. First case study was inspired by an artistic decorative floral pattern. Second case study describes the detailed process of developing a new L-system grammar that leads to the original graphics.

Keywords: Formal grammar · L-system · Generative art · Evolutionary art · Genetic programming

1 Introduction

Lindenmayer systems (L-systems) are formal grammars with parallel rewriting mechanism that were originally developed for modelling and visualization of the growth process of various types of algae [17]. Later they were applied in the field of computer graphics. The best known graphical interpretation of L-systems is based on usage of a relative cursor upon a Cartesian plane (turtle graphics). The L-systems are frequently used in combination with evolutionary techniques, e.g. [9] presents a parametric L-system for drawing virtual creatures for computer animations.

In this contribution we explore the emergence phenomena growing from combination of L-systems, genetic programming and interactive evolution, primarily we are interested in its application in generative art and artificial creativity. We were inspired by shape grammars that also originated from the theory of formal grammars: it was shown that even simple rules of shape grammars produce complex results [22] moreover [8] applied shape grammars in design.

The key issue in our research area is the definition of the fitness function. Artificial neural networks or design principles measurements are well-applicable techniques. Methods that do provide automatic fitness function for aesthetic evaluation are

© Springer-Verlag Berlin Heidelberg 2016
N.T. Nguyen et al. (Eds.): TCCI XXIV, LNCS 9770, pp. 83–100, 2016.
DOI: 10.1007/978-3-662-53525-7_5

classified as Computational Aesthetic Evaluation (EAC) systems. For relevant results see e.g. [1, 2, 18, 19].

In our case we chose opposite approach: the fitness of the individuals is assigned manually by the user. The approach is known as interactive evolutionary computation (IEC) [7].

The graphical interpretation of L-system strongly depends on its definition, because even minor changes of parameters lead to completely different and surprising results. The evolutionary techniques help to search the large space of parameters and to modify production rules.

Current implementations of the L-system theory are based on extensions of general purpose programming languages. From the technical point of view, languages such as L+C [12], L-Py [4] or XL [13] are complex and their implementations are platform dependent. Our intention is to provide easy-to-use tool for partly interactive creation of various types of graphical outputs. On the contrary, our tool does not make user to directly write production rules, in fact the user does not need to know the grammar of the L-system at all.

In following sections of this paper we propose a new formal language (L2 language) which is easy to parse to the tree representation. Then we provide a platform independent tool (L2 Designer) which enables specification of L-systems within L2 language, with its subsequent evolution based on genetic programming. Finally, the graphical interpretation of outputs is presented.

2   L2 Language and L-system Extensions

L2 language is designed for defining stochastic context free parametric L-systems grammars. In contrast to L+C or L-py, L2 does not include anything else but features that we need for the purpose of the definition of L-system. As well as L+C, L2 supports the advanced properties of L-systems:

- **Sub-L-systems** – it is possible to divide large L-systems into smaller reusable parts,
- **Interpretation production rules** – it lets us separate topology of the L-system from its **representation**, therefore the application of genetic programming operators is easier.

For detail specification of L2 language, see the documentation [14]. Here we provide a sample code. Its explanation and interpretation is shown below.

As shown in Fig. 1, the L2 program consists of three main parts:

- **Alphabet** - a set of symbols (line 1),
- **L-script** encapsulating L-systems (line 5),
- **L-system** - the unit defining production rules, default axiom and default number of derivations (lines 6, 16).

Both the variable and parameter names start with a symbol $. The L-script contains the main L-system (line 29).

A *derive* statement (line 31) launches the derivation process of the L-script, whose identifier is passed as the argument.

```
01. alphabet Turtle2D {
02.    F, f, L, R, PU, PS
03. };
04. $black = __rgb(0,0,0,255);
05. lscript BranchingLScript {
06.    lsystem Bloom(F(0.01), 3) using Turtle2D {
07.      $angle = 90;
08.      $colorA = __rgb(255,100,0,200);
09.      $colorB = __rgb(150,50,50,200);
10.      F(a) --> F($a) L($angle) F($a) A($a);
11.      F(a) --> F($a) R($angle) F($a) A($a);
12.      F(a) -h> F($a, 0.003, __rgb(0,0,0,0));
13.      A(a) -h>
14.       [ F(0.0001, $a * 1.5 * __random(), $colorA) ]
15.       | [ F(0.0001, $a * 1.5 * __random(), $colorB) ];
16.    };
17.    lsystem Branching(G(0.1), 4) using Turtle2D {
18.      $ratio = 0.9;
19.      $angle = 60;
20.      $anglePrec = 50;
21.      $stroke = 0.003;
22.      G(a) --> F($a) [ L($angle) ($ratio * $a)B($a) ]
23.       [ R($angle) G($ratio * $a) B($a) ];
24.      L(a) -h> L($a - $anglePrec * 0.5);
25.      R(a) -h> R($a - $anglePrec * 0.5);
26.      F(a) -h> F($a, $stroke, $black)
27.              [sublsystem Bloom(F($a / 10), 6)];
28.    };
29.    main call Branching();
30. };
31. derive BranchingLScript;
```

Fig. 1. L2 sample code

BranchingLScript contains a definition of two L-systems called *Bloom* and *Branching* (lines 6–16, 17–28). The heading of the L-system consists of:

- name,
- default axiom,
- default number of derivations,
- alphabet.

The body of L-system contains the list of productions. For the L-system productions we use either the —> operator (line 10) or the $-h >$ for interpretation rules (line 12). The replacement string on the right side of the production rule can contain *sublsystem* statement (line 27) that calls derivation of the other L-system within the same L-script.

3 L2JS Library

The L2 language is accessible within our L2JS library which is the core of L2 Designer. The library includes compiler, interpreter and module for genetic programming.

3.1 Compilation Process

The compiler of the L2JS library translates L2 to JavaScript. The scripting language was chosen due to its flexibility, dynamic scoping, closures and both functional and object-oriented programming support.

The usage of JavaScript allows us to distribute the computing within the web browser and Node.js web server.

Figure 2 shows the flow of the compiling process. The compilation starts with the linking of the input source files. The whole code is parsed by Jison parser [11] into the abstract syntax tree (AST). The AST consists of the nodes representing statements, arguments, entities, names, variables and expressions. Jison is a JavaScript implementation of the combination of [5] and [3]. The parser requires the L2 grammar description file. After the L2 AST is created by Jison, the translation to the JavaScript code can be performed.

Our compiler is able to decompile L2 AST back to the L2 code. This feature is essential for further application of genetic programming.

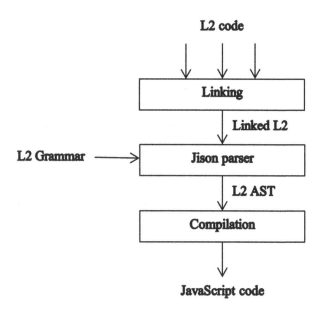

Fig. 2. Scheme of the L2JS library

The output of the compilation process is a JavaScript program representing the derivation of the L-script.

3.2   Interpretation

The Interpreter operates with an alphabet of the L-system to resolve the type of interpretation. In our sample code, an alphabet Turtle2D is used in which the symbols of the alphabet are understood as the instructions for the turtle graphics (Fig. 3). The interpretation of the symbols is drawn to the HTML5 Canvas directly inside the browser. For the description of the symbols see [20].

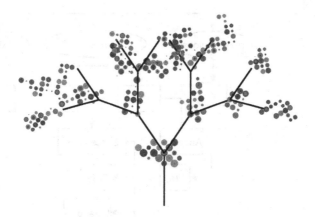

Fig. 3. Turtle graphics for the sample code

The Interpreter works with rules that specify how symbols are related to the set of statements from the alphabet. E.g. the module *F(a)* is replaced with *F($a, 0.003, __rgb (0,0,0,0))* according to the corresponding interpretation rule (line 12 in Fig. 1). The interpretation in this particular case results in a-long line with 0.003 size filled with transparent colour.

The Evolver implements the L-system genetic programming over the hierarchy of L-systems in L2 language. Details are provided in the next section.

4   Genetic Programming

Genetic programming is involved in the process of iterative modifications of L-scripts.

After the initial L-script is provided to the Evolver module, it becomes the base for the initial generation. Each individual is represented by the L-script which is converted to L2 AST (see Fig. 4 for illustration). The user has to specify which of sub-L-systems should be modified by the Evolver.

To represent the genotype in the Evolver module we decided to use a tree data structure, as it was originally introduced by Koza in [15]. Genetic operators modify the

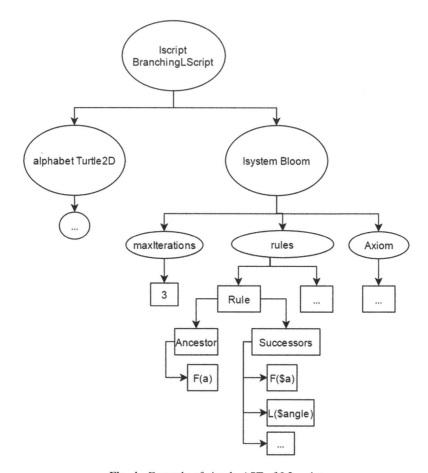

Fig. 4. Example of simple AST of L2 script

abstract syntax trees using crossover or mutation. The genetic operators designed for the parametric L-systems were originally proposed by [19].

For chosen types of the AST nodes, we are using different operators. The operators are applied either on the L-systems (axiom, production rules) or on the expressions used within the program (variable assignments, expressions within modules). The Evolver module supports:

- mutation of production rules,
- crossover of production rules,
- mutation of expressions – variation, creation and colour mutation.

The main task was to identify the right terminals. We found the way of an automatic detection of terminals without the need of its explicit specification by the user.

In the case of production rule mutation, the set of terminals consists of distinct symbols occurring in all production rules of parent L-system.

The set of terminals depends on the context of expression in the case of expression mutation. On the other hand the terminals within the rule are enriched with parameters from the ancestor of the rule.

The Evolver implements two types of mutations. The first of them modifies the numbers only (variation mutation), the second one generates new expressions (creative mutation).

Special mutation operator was developed for the colours: user can specify maximum percentage of modification for each of the channel of the colour model. For hue channel it is possible to specify an exact angle that can be added or reduced.

We are using a tree representation of the production rules similarly to [10]. Every leaf is a module of the rule and every new level of the tree is determined by stack symbols ([,]).

For example representation of the L-system string $F [G [H] I [J K]]$ is shown in the Fig. 5.

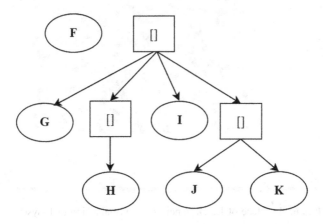

Fig. 5. Tree representation of the string containing nested stack symbols

The crossover is represented by an exchange of the branches from the tree representation of L-systems. The newly created production rule either replaces the rule which was selected for the crossover (parent rule) or is added to the definition of the L-system. Probability of the newly created rule is determined by product of the probability of parent rule and a predefined constant.

In the process of the fitness evaluation, the user selects the best solutions generated by the program and assigns the integer fitness values to these solutions. In comparison with other methods, IEC is more time consuming, only small populations and low number of generations can be processed effectively. On the other hand, with IEC the user can apply his aesthetic preferences.

The linear rank selection mechanism is combined with elitism. For details of this method see [21].

5   Implementation

The L2 Designer is a web-based JavaScript application enabling the interactive designing of L-systems. The core of the application is L2JS library. The server is running on Node.js. Other main technologies we are using are MongoDB and Angular.js.

Within the L2 Designer the user can manage projects and directories of scripts. The main focus is on the process of designing new L-systems (Fig. 6). The source codes are available together with their interpretations.

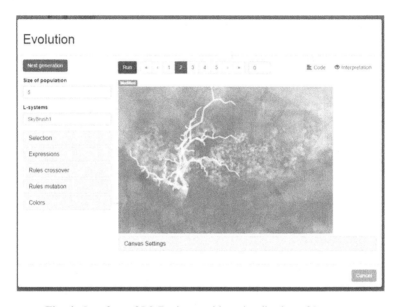

Fig. 6. Interface of L2 Designer with a visualization of L-system

6   Case Studies

Let us demonstrate the L2 Designer workflow. We prepared two case studies of L-systems generating complex images. The first one is meant to be visually similar to already existing art piece; however the second case study is intended to create the original graphics.

6.1   Michaelmas Daisy

Let us demonstrate the L2 Designer workflow. The decorative floral pattern Michaelmas Daisy 1929 (Fig. 10) was our inspiration.

The process starts with creation of L-script which contains several sub-L-systems. The first part of L-systems represents basic shapes (flower petal, leaf, disc floret); the

Fig. 7. Layout of leaves before the interactive evolution computation: completely regular distribution of leaves of one size, with limited number of colour shades (Color figure online)

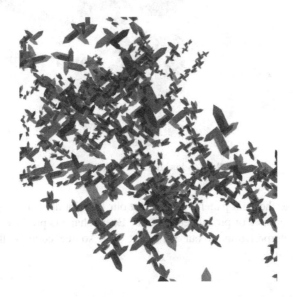

Fig. 8. Layout of leaves after the interactive evolution computation: irregular distribution of leaves on the canvas, higher number of colour shades and variable sizes (Color figure online)

second part represents a layout of basic shapes (flower head, layout of flowers, layout of leaves).

The aim of interactive evolutionary computation is to increase the similarity of the output graphic interpretation of L-system with the original pattern. The original pattern background is covered with leaves. The interpretation of the L-system before and after the process of IEC is shown in Figs. 7 and 8 respectively. The next step is the creation of the main L-system which generates the layout of flowers. First of all it is necessary to define L-system for random distribution of flower heads over the canvas. This initial result still does not correspond to the original artifact (Fig. 9): there is a lack of the grouping of flowers of the same type. Again, this issue can be solved by evolution of the L-system.

Fig. 9. Random distribution of flowers

See Fig. 10 for the final pattern. Notice that our L-script does not cope in any way with external image files or predefined patterns. Every shape is produced solely by the turtle graphics interpretation of our L2 script. The source code of the L-script is available online [6].

6.2 Generative Scenery

For the second example we decided to develop the L-system without explicit reference to already known visual art piece. Originally the L-system was intended to represent the abstract drawing of the landscape containing the cloudy sky and the grassy surface.

Sky. Starting with the sky we searched for the sub-L-system, which creates simple 2D artifacts. Such L-system was meant to be used by the parental L-system (evolved

Fig. 10. Final pattern resulting from the evolution of L-system (left) inspired by Michaelmas Daisy (right) [23]

Hilbert curve) to cover the canvas. However during the process of evolving the sub-L-system we found that the combination of different individuals is sufficient enough to make the drawing of the sky completed without need of usage the parental Hilbert curve.

Respectively Figs. 11 and 12 show both the interpretation and L2 code for the original spiral L-system. The simple L-system contains only one rule for each type of production rules.

Each derivation adds the symbol C with increased parameter f by the constant inc to the resulting string. Symbol C is interpreted with the corresponding interpretation rule as the circle with the radius f, filled with the colour represented by HSV model.

Fig. 11. Original spiral L-system

```
$skyHue = 200;
$skyAlpha = 50;
lsystem SpiralCircle(A(0.04), 6) using Turtle2D {
  $angle = 60;
  $inc = 1.1;
  A(f) --> [ f(2 * $f) C($f) ] L($angle) A($f * $inc);
  C(f) -h> C($f, __hsv($skyHue,1,1,$skyAlpha));
};
```

Fig. 12. L2 code for the spiral L-system

Table 1. Settings for the evolution process of the spiral L-system

Parameter	Value
Population size	100
Elite individuals	10
Probability of expression mutation	0.5
Probability of rule crossover	0.8
Probability of rule mutation	0.5
Probability of adding the rule as a new rule (rule created by crossover operator)	0.3
Probability of adding the rule as a new rule (rule created by mutation operator)	0.3
Colour variation of the H/S/V channels	5 %/5 %/5 %
Set of angles that can be added to the H-channel when applying the colour mutation	{60°, 30°, 90°}

The spiral L-system determined the individual for the first generation of the evolving process. Table 1 shows the summary of parameters used during the evolution process. Moreover the settings were used for the design of all L-systems in this case study.

Selection of several interesting individuals shows the Fig. 13. The last example from the figure shows the pattern which sufficiently covered the canvas to be considered as the base L-system for the next evolution process. For the code of L-system SpiralSkyBase see Fig. 14. Our Evolver made changes to both constants $angel and $inc. Next changes were made in the main production rule for the symbol A.

At the first sight, the evolved expressions are unnecessarily complex. E.g. the *$f * $f - $f - $f + $f + ($f))* can be easily simplified to *$f * $f*. However the complexity of the expression is important for our implementation of the crossover operator. Due to larger expression tree there is higher probability of creative permutations in the next generations. The maximum level of the expression tree can be explicitly set to our Evolver module.

In the evolution of L-system *SpiralSkyBase* we searched for the colourful individuals as well as for the patterns preserving the central composition. We picked several of the individuals to compose the drawing of the sky. The Fig. 15 shows each

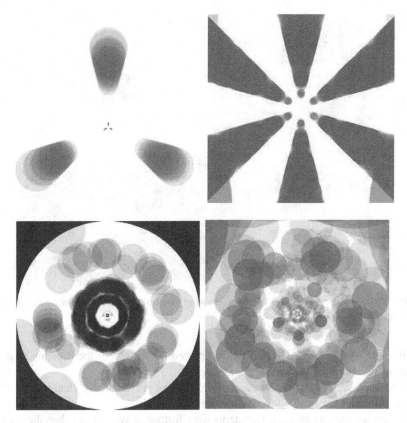

Fig. 13. Sample of individuals from evolution of spiral L-system

```
lsystem SpiralSkyBase(A(0.04), 6) using Turtle2D {
  $angle = 70.23168000000001;
  $inc = 1.1891;
  A(f) -->
    [ f(2 * $f) C($f) ] A($f + $f + ($f))
    [ A($f) A($f * $f - $f - $f + $f + ($f)) ]
    L($angle) A($f * $inc);
  C(f) -h> C($f, __hsv($skyHue,1,1,$skyAlpha));
};
```

Fig. 14. L2 code for *SpiralSkyBase*

of them separately. During the process we found L-system *Sun* that is placing circles in different sizes on top of each other. The L-system suits well as the sun for our drawing. The L-system *SkyBackground* is a denser stochastic version of the original L-system *SpiralSkyBase*. To add more colours we picked the stochastic L-system *SkyHue*.

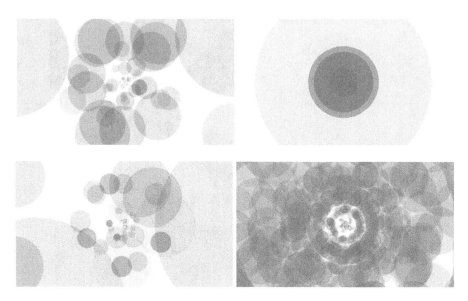

Fig. 15. Picked individuals for the sky drawing in order: *SkyReflection*, *Sun*, *SkyHue*, *SkyBackground* (Color figure online)

Finally the L-system *SkyReflection* will be placed on the top of the drawing to make feeling of reflection of the sun in the landscape.

In both of the L-systems *Sun* and *SunReflection* we manually changed the *$hue* constant to make the sun yellow and the reflection white.

We can see that the evolved L-systems were further more complex than the original one - according to both grammar and visual interpretation.

Countryside. To add the actual land into our drawing, we chose to evolve Hilbert curve. Such L-system is going to place 2D artifacts in certain spots to make the texture of the countryside. Hilbert curve was well suited as the member of Space-filling curves

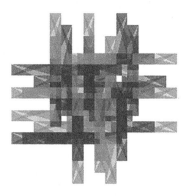

Fig. 16. Base L-system *ColoredHilbertCurve*

```
$grassHue = 10;
$grassAlpha = 100;
lsystem ColoredHilbertCurve(A(0.01), 3) using Turtle2D {
  $angle = 90;
  A(f) --> L($angle) B($f) F($f) R($angle) A($f) F($f)
       A($f) R($angle) F($f) B($f) L($angle);
  B(f) --> R($angle) A($f) F($f) L($angle) B($f) F($f)
            B($f) L($angle) F($f) A($f) R($angle);
  F(f) -h> f($f)
          [ F($f * 4, $f,
          __hsv($grassHue,1,1,$grassAlpha)) ];
  F(f) -h> f($f)
          [ F($f * 4, $f,
          __hsv($grassHue + 50,1,1,$grassAlpha)) ];
};
```

Fig. 17. L2 code for the *ColoredHilbertCurve* (Color figure online)

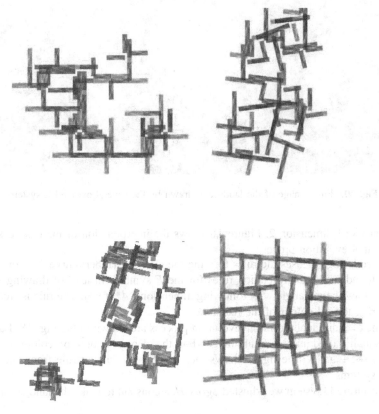

Fig. 18. Sample of individuals from the evolution of *ColoredHilbertCurve*. Last two were selected for the drawing of the countryside. (Color figure online)

```
F(f)  -h> F($f * 4, $f,
            __hsv(
                $grassHue + 44.13,
                0.941,
                0.828,
                1.067 * $grassAlpha)
        )
        [ F($f * 4, $f,
            __hsv($grassHue + 50,1,1,$grassAlpha)) ];
```

Fig. 19. Example of evolved complex production rule (Color figure online)

Fig. 20. Final image of the landscape drawn by the several evolved L-systems

with Hausdorff dimension 2. Figure 16 shows the interpretation of the base L-system for our final evolution process.

The code Fig. 17 uses the rules forming the classical Hilbert curve. The difference can be found in the interpretation rules for the *F* symbol. Instead of drawing simple line, we draw the rectangle with following dimensions: *4*$f x $f*. The rule is stochastic considering the fill colour of the rectangle.

Some of individuals from the evolution process are shown in the Fig. 18. Last two were actually used in the final image. For the demonstration of colour mutation operator see Fig. 19. The figure shows the complex production rule from one of the final L-systems.

In resulting L-system we adjusted *$grassHue* constant to make the land coloured in blue.

Composition. Finally, we wrote parental L-system that composes our evolved L-systems into the final drawing (Fig. 20). Stochastic nature of L-system allows generating different images by each derivation (Fig. 21). The complete source code and the animation from the interpretation can be found in [16].

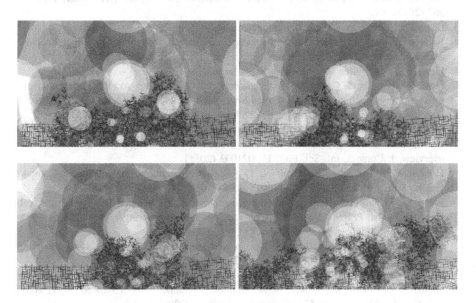

Fig. 21. Stochastic nature of the final L-system

7   Conclusions

The current version of L2 Designer is capable of evolving stochastic parametric L-systems which are described by L2 language and generate complex graphical patterns. The proposed case studies shows that evolving simple L-systems grammars by our Evolver module can lead to both complex grammars and interesting graphical interpretations.

Our evolutionary process is strongly influenced by user preference; the results are generally unique for every new process of the design.

Our next intention is to improve the effectiveness of genetic operators to speed up the fitness evaluation. For that we are going to implement a module for the processing of user's input using graphic tablet or vector image file.

The development of L2 language continues. New features will be added such as decomposition rules or rule conditions.

Finally, with respect to the generative art applications, we want to explore the possibility of integration of L2 Designer with graphical editors to support creativity in users.

Acknowledgements. Excellence Project "MAS Applications in Modeling of Complex Socioeconomic Systems and Intelligent Environments" is gratefully acknowledged.

References

1. Ashlock, D., Bryden, K.M.: Evolutionary control of Lsystem interpretation. In: CEC 2004, vol. 2, pp. 2273–2279 (2004)
2. Bergen, S., Ross, B.J.: Aesthetic 3D model evolution. Genet. Program. Evol. Mach. **14**, 339–367 (2013)
3. Bison (2015). http://www.gnu.org/software/bison/
4. Boudon, F., et al.: L-Py: an L-system simulation framework for modeling plant architecture development based on a dynamic language. Front. Plant Sci. **3**, 76 (2012)
5. Flex (2015). http://flex.sourceforge.net/
6. Flower Pattern Source Code [WWW Document], n.d. GitHub. https://raw.githubusercontent.com/tommmyy/l2js/master/app/js/lscripts/flowerpatternevolved.l2. Accessed 20 Feb 2016
7. Galanter, P.: Computational aesthetic evaluation: past and future. In: McCormack, J., d'Inverno, M. (eds.) Computers and Creativity, pp. 255–293. Springer, Heidelberg (2012)
8. Chakrabarti, A., Shea, K., Stone, R., et al.: Computer-based design synthesis research: an overview. J. Comput. Inf. Sci. Eng. **11**, 021003 (2011)
9. Hornby, G.S., Pollack, J.B.: Evolving L-systems to generate virtual creatures. Comput. Graph. Artif. Life **25**, 1041–1048 (2001). doi:10.1016/S0097-8493(01)00157-1
10. Jacob, C.: Genetic L-system programming. In: Davidor, Y., Männer, Reinhard, Schwefel, Hans-Paul (eds.) PPSN 1994. LNCS, vol. 866, pp. 333–343. Springer, Heidelberg (1994)
11. Jison (2015). http://zaach.github.io/jison/
12. Karwowski, R., Prusinkiewicz, P.: Design and implementation of the L+C modeling language. Electron. Notes Theor. Comput. Sci. **86**(2), 134–152 (2003)
13. Kniemeyer, Ole, Kurth, Winfried: The modelling platform GroIMP and the programming language XL. In: Schürr, Andy, Nagl, Manfred, Zündorf, Albert (eds.) AGTIVE 2007. LNCS, vol. 5088, pp. 570–572. Springer, Heidelberg (2008)
14. Konrády, T.: L2 documentation (2015). https://github.com/tommmyy/l2js/wiki/Documentation. Accessed 20 Feb 2016
15. Koza, J.R.: Genetic Programming. 1: On the Programming of Computers by Means of Natural Selection. MIT Press, Cambridge (2000)
16. Landscape Source Code [WWW Document], n.d. GitHub. https://github.com/tommmyy/l2js/tree/master/app/js/lscripts/landscape. Accessed 1 Jan 2016
17. Lindenmayer, A.: Mathematical models for cellular interactions in development. J. Theor. Biol. **18**, 280–315 (1968). Elsevier, Part I and II
18. McCormack, J.: The application of L-systems and developmental models to computer art, animation and music synthesis (2003). http://www.csse.monash.edu.au/~jonmc/research/thesis.html
19. McCormack, J.: Evolutionary L-systems. In: Hingston, P.F., Barone, L.C., Michalewicz, Z. (eds.) Design by Evolution, Natural Computing Series, pp. 169–196. Springer, Heidelberg (2008)
20. Node.js (2015). https://nodejs.org/
21. Sivaraj, R., Ravichandran, T.: A review of selection methods in genetic algorithm. Int. J. Eng. Sci. Technol. **3**, 3792–3797 (2011)
22. Stiny, G.: Shape rules: closure, continuity, and emergence. Environ. Plan. **21**, 49–78 (1994)
23. The Warner Textile Archive (2015). http://www.warnertextilearchive.co.uk/

Manifold Learning Approach Toward Constructing State Representation for Robot Motion Generation

Yuichi Kobayashi$^{(\boxtimes)}$ and Ryosuke Matsui

Department of Mechanical Engineering, Shizuoka University,
3-5-1 Johoku, Naka-ku, Hamamatsu, Shizuoka, Japan
kobayashi.yuichi@shizuoka.ac.jp
http://www.sensor.eng.shizuoka.ac.jp/~koba

Abstract. This paper presents a bottom-up approach to building internal representation of an autonomous robot. The robot creates its state space for planning and generating actions adaptively based on collected information of image features without pre-programmed physical model of the world. For this purpose, image-feature-based state space construction method is proposed using manifold learning approach. The visual feature is extracted from sample images by SIFT (scale invariant feature transform). SOM (Self Organizing Map) is introduced to find appropriate labels of image features throughout images with different configurations of robot. The vector of visual feature points mapped to low dimensional space express relation between the robot and its environment with LLE (locally linear embedding). The proposed method was evaluated by experiment with a humanoid robot collision classification and motion generation in an obstacle avoidance task.

Keywords: Developmental robotics · Humanoid robot · Manifold learning · Image features

1 Introduction

One of the difficulties which autonomous robots face in non-structured environment is that they are not ready to unexpected factors and changes of their environments. In actual applications, it is not robots themselves but human designers or operators that detect, analyze and find solutions for the unexpected factors. In other words, adaptability of autonomous robots with current technologies is not sufficient as to let them act in environments close to our daily life. One promising approach to overcome the lack of adaptability of autonomous robots is to build intelligence of robots in a bottom-up manner, known as developmental robotics [13] and autonomous mental development [20]. They have a common idea for building robot intelligence, e.g., stress on embodiment, self-verification [17], mimicking developmental process of human (infant) [15], etc.

© Springer-Verlag Berlin Heidelberg 2016
N.T. Nguyen et al. (Eds.): TCCI XXIV, LNCS 9770, pp. 101–116, 2016.
DOI: 10.1007/978-3-662-53525-7_6

Among various concerns in the field of developmental robotics, problem of building state space, with which a robot can plan and control its action, is rather important but has not been gathering sufficient attention. One reason for this is that imitation learning, generating appropriate robot motions based on human demonstration [2,3], is much more effective to generate complex motions with high degrees of freedom. It is known that acquisition of motion without any pre-defined knowledge on robot tasks, *e.g.*, by reinforcement learning [18,19], takes numerous trials. Thus, it is not directly applicable to continuous high-dimensional control problems, except for some cases where motions of robots are restricted to continuous trajectory generation without interacting with objects (*e.g.*, [4,5]). The problem of constructing state space, however, is remaining to be of great importance for autonomous robots to finally generate, control and modify their motions adaptively, even though prototype motion could be built by imitation initially.

Generation of space which is suitable for robot motion learning has been investigated from various viewpoints. One example of space construction is related to visual attention [6], where sequences of successful motions had been provided with robot in advance. Poincaré map is another example of abstract representation for complex robotic behavior learning [14], where periodic walking pattern by a biped robot was considered.

Apart from researches on acquisition of behavior of robot itself, such as walking, jumping, and standing up, state space construction has not been regarded as an important issue. In general, configurations of objects and robots are assumed to be observable in researches on manipulating objects, where positions and postures of objects in the Cartesian (world) coordinate system are used as a solid base.

But in the real world application, measurement of 3D configurations of objects is difficult. It contains difficulties in multiple levels:

1. The framework of 3D configuration measurement inherently requires measuring precise shape of an object, but it is difficult to measure whole shape of an object because measurement by camera or laser scanner is normally unilateral.
2. Spatial relation between robot and object is generally very important for both object manipulation and collision avoidance, whereas an object is more likely to be occluded by the robot when the robot is approaching to the object.
3. Deformation of object is normally not considered or requires specific model for mathematical analysis. But it is difficult to precisely model the deformation.

From the viewpoint of the developmental robotics, the 3D representation in the world coordinate is not a sole way to express the state for a robot. If a robot can build representation of its environment based only on what it can verify by itself, the representation might not suffer from the above-mentioned difficulties (as can be seen in a learning approach [7]).

This paper presents an approach to the interest of *building a representation of a robot* for motion planning and control *in an adaptive way without any pre-defined knowledge*. To consider relation between the robot and its environment, image features based on SIFT (Scale Invariant Feature Transform) [12] are used. Image feature-based learning of robot behavior was presented in [9], but it did not deal with relation between an object and the robot with a quantitative representation. In this paper, application of a manifold learning is introduced, which enables not only to classify state of the robot but also to evaluate how much the robot is close to a certain state. In addition, the obtained representation will be utilized for motion generation of collision avoidance.

Locally Linear Embedding (LLE) [16] is used as a means for manifold learning because continuous property of the system can hold only in a local region in the problem of recognition of environment by a robot. For the application of LLE, vector generation based on SIFT-features matching is proposed to deal with a problem that keypoints of SIFT are not consistent through all the images. The proposed framework is evaluated in experiment using a humanoid robot, preceded by preliminary verification of LLE framework with simulated image vectors.

The remainder of the paper is organized as follows. The problem settings are described in Sect. 2. The proposed method of constructing state representation including its application to motion generation is described in Sect. 3.

2   Problem Settings

Images obtained by CCD camera attached at the head of a robot are considered as input to the robot system, as indicated in Fig. 1. Humanoid robot NAO [1] is considered both in simulation and experiment. The images contains part of body of the robot, an object which can contact with the robot's body and background which are not affected by configuration of the robot. The configuration of the robot arm is shown in Fig. 2. Shoulder roll joint and shoulder pitch joint are controlled (ϕ_1 and ϕ_2), while other two joints are fixed throughout the experiment. This implies that the motion of the robot arm is constrained on a plane which is vertical to optical axis of the CCD camera. A red plane in Fig. 1 is parallel to the motion constraint plane.

Image features are extracted from each image, as shown as circles in the right hand of Fig. 1. Keypoints of SIFT [12] are used as image features. The robot does NOT have knowledge on properties of image features, that is, the robot does not have labels of what is object or what is robot's body in the image in advance. The robot collects images while changing configuration of its arm. Position of the object can also differ irrelevantly to the position of the arm.

Objective of the robot system is to construct a space which provides the following functions:

1. Estimating closeness of its hand to the object
2. Predicting collision of its hand with the object

The first function allows the robot to plan its hand trajectory so as not to be too close to the object, when the robot intends to achieve a task while avoiding collision with obstacles. The second function does not directly allows the

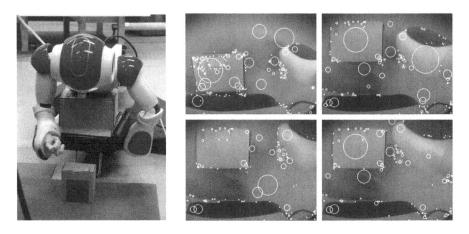

Fig. 1. Humanoid robot NAO and its image (Color figure online)

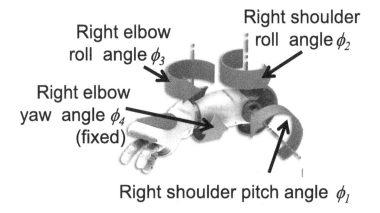

Fig. 2. Configuration of robot arm (right)

robot to avoid collision, but can contribute to the ability by integration of other techniques, *e.g.*, prediction of robot's hand in the image space.

3 Manifold Learning Using Image Features

Manifold learning by LLE is applied to the SIFT keypoints to obtain a continuous space which reflect relation between the hand and its environment. Each keypoint has 128-dimensional feature vector that can be utilized for identification and matching to other keypoints. By the matching process, a keypoint can be traced through multiple images if it is extracted commonly in those images. One problem in generating a vector for manifold learning is that feature vector of a keypoint is not consistent in different images due to change of posture of the arm. The arm, which consists of serial links, inevitably change its posture even when the end of the arm is making translation. Under an assumption that each keypoint tracks a certain part of the arm, a method for matching and labeling is proposed using Self Organizing Map (SOM) [11].

Figure 3 shows an overview of the proposed state space construction approach. Images captured by the camera at the head of the robot are source of the process. SIFT keypoints provide vectors which express positions of feature points in the images. A low-dimensional space is constructed based on the vector by using LLE.

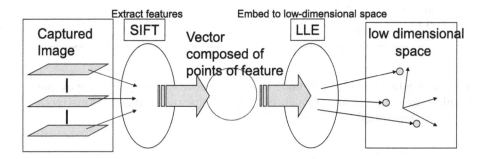

Fig. 3. Overview of state space construction

Hereafter, D denote dimension of image vector, N denote total number of images and $I^{(i)} \in \mathbb{R}^D, i = 1, \cdots, N$ denote vector of image i. $M(i)$ denote number of keypoints in image i.

3.1   Matching and Labeling of Features

First, image vectors $I^{(i)}, i = 1, \cdots, N$ are used to generate a SOM. Let K denotes total number of nodes in SOM. Image vectors are divided into sets by the nodes of the SOM.

$$G(k) = \{i | k = \arg \min_\ell \|I^{(i)} - \boldsymbol{w}_\ell\|^2\}, \tag{3.1}$$

where $\boldsymbol{w}_k \in \mathbb{R}^D$ denotes weight vector of node k. $G(k)$ denotes set of images that are similar to $\boldsymbol{w}_k$. For each node, a representative image is decided as

$$\bar{i}_k = \arg \min_{i \in G(k)} \|\boldsymbol{w}_k - I^{(i)}\|^2, \ k = 1, \cdots, K. \tag{3.2}$$

Image $\bar{i}_k$ is used for generating labels of keypoints. Labels are generated by Algorithm 1. As a sequel to the labeling procedure, totally $\sum_{k=1}^K M(\bar{i}_k)$ labels are generated.

Although feature vector of a keypoint can differ by the change of the robot's configuration, it is likely that feature vectors in images with small differences are similar. By using topological neighbor of SOM, correspondence between keypoint labels can be found. Figure 4 indicates the idea of combining redundant labels.

Algorithm 1. Labeling of keypoints

---

  for $k = 1$ to K **do**
    Select representative image $\bar{i}_k$ for node k
    for $\ell = 1$ to $M(\bar{i}_k)$ **do**
      Select keypoint ℓ in image $\bar{i}_k$
      for $i = 1$ to N **do**
        if $i \notin G(k)$ **then**
          Apply SIFT matching with keypoint ℓ to all keypoints in image i
          If matching found, label it
        end if
      end for
    end for
  end for

For representative node $\bar{i}_k$ in node k, feature vectors of keypoints are averaged within matched keypoints of images $i \in G(k)$. Using the averaged feature vectors, labels are integrated by Algorithm 2.

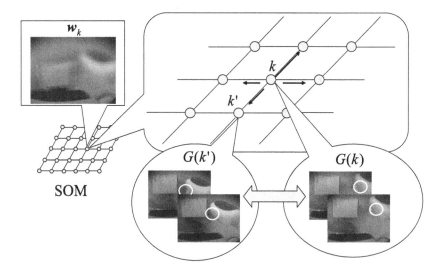

Fig. 4. Matching of image features

By finding correspondence between neighbor nodes, labels which correspond to the same part of the real world are integrated into one label.

3.2   Space Construction with LLE

Using the obtained labels in the previous section, vectors are defined as follows. Let L denote the number of integrated labels. Keypoint information of image i

---

Algorithm 2. Integration of labels

---

for $k = 1$ to K **do**

   Find neighbor nodes of node k as $i' \in \mathcal{N}(k)$

   for $\ell = 1$ to $M(\bar{i}_k)$ **do**

     for $i' = 1$ to $|\mathcal{N}(k)|$ **do**

       Apply SIFT matching with keypoint ℓ by average feature vector

       If matching found, record correspondence between ℓ and the matched label

     end for

     If no matching found, remove label ℓ

   end for

end for

Integrate all labels using recorded correspondence

---

is converted to vector $\boldsymbol{x}_i \in \mathbb{R}^{2L}$, where $\boldsymbol{x}_i$ is defined by

$$\boldsymbol{x}_i = [u_1^{(i)}\ v_1^{(i)}\ u_2^{(i)}\ v_2^{(i)}\ \cdots\ u_L^{(i)}\ v_L^{(i)}]^T. \tag{3.3}$$

$(u_\ell^{(i)}, v_\ell^{(i)})$ denotes position (image coordinate) of keypoint whose label is ℓ in image i. If keypoint whose label is ℓ does not exist in image i, averages over all images are used for $(u_\ell^{(i)}, v_\ell^{(i)})$. Finally, data matrix for LLE is constructed as

$$H = [\boldsymbol{x}_1\ \boldsymbol{x}_2\ \cdots\ \boldsymbol{x}_N] \in \mathbb{R}^{2L \times N}. \tag{3.4}$$

LLE is a method which maps a high-dimensional vector ($2L$ in this application) to a low-dimensional vector while preserving local linear structure of each data around its neighborhood. Weighting coefficient $\boldsymbol{v}_j^i, j = 1, \cdots, n$ for sample i, where n denotes the number of neighborhood, is calculated so that the cost function defined by the following is minimized.

$$\epsilon_1 = \sum_{i=1}^{N} \| \boldsymbol{x}_i - \sum_{j=1}^{n} v_j^i \boldsymbol{x}_j^i \|^2, \tag{3.5}$$

where $\boldsymbol{x}_j^i$ denotes neighborhood sample of $\boldsymbol{x}_i$. A low-dimensional vector $\boldsymbol{y}_i \in \mathbb{R}^d$, corresponding to $\boldsymbol{x}_i$, is calculated so that the following cost function is minimized.

$$\epsilon_2 = \sum_{i=1}^{N} \| \boldsymbol{y}_i - \sum_{j=1}^{n} v_j^i \boldsymbol{y}_j^i \|^2, \tag{3.6}$$

where $\boldsymbol{y}_j^i, j = 1, \cdots, d$ denotes neighborhood of $\boldsymbol{y}_i$ and d denotes dimension of the low-dimensional space.

3.3   Motion Generation

Dynamic programming with discrete state representation [18] is applied for motion generation. The state for motion generation is defined by the joint angle space. The discrete state $s \in \mathcal{S}$ is given by discretising the joint angles of the robot (ϕ_1, ϕ_2) into $N_{s1} \times N_{s2}$ grids, where $\mathcal{S}$ denotes set of states and N_{s1} and N_{s2} denote grid size for ϕ_1 and ϕ_2, respectively. Action $a \in \mathcal{A}$ is defined as four directional transitions from a grid to its adjacent grids, where $\mathcal{A}$ denotes set of actions. Let $G \subset \mathcal{S}$ denote set of goal states. The reward $r(s, a)$ for $s \in \mathcal{S}$ and $a \in \mathcal{A}$ is given by the following:

$$r(s, a) = \begin{cases} 0, & \text{if } s' \in G \\ -100, & \text{if collision happens} \\ -1, & \text{otherwise,} \end{cases} \tag{3.7}$$

where s' denotes the next state reached by action a from state s. Using this reward setting, motion with minimum time (action-step) can be obtained through the value iteration.

The value function $V(s)$ is calculated by the Bellman equation with deterministic state transition:

$$V(s) = \max_{a \in \mathcal{A}} [r(s, a) + \gamma V(s')], \; {}^{\forall}s \in \mathcal{S}, \tag{3.8}$$

where s' denotes the next state after transition, $\gamma (0 < \gamma \leq 1)$ denotes the discount factor. The action value function $Q(s, a)$ is given by the following:

$$Q(s, a) = [r(s, a) + \gamma V(s')]. \tag{3.9}$$

At every time step, the robot decides its action using the following the obtained policy $a = \pi(s)$, which is given by the following:

$$\pi(s) = \arg \max_{a \in \mathcal{A}} Q(s, a) \tag{3.10}$$

Based on the policy, the robot moves its joint angles (ϕ_1, ϕ_2) toward its adjacent grid value.

4   Experiment

The proposed representation was evaluated by experiment in two ways, with simulated images and actual images obtained by a CCD camera attached at the head of the robot. A motion generation of collision avoidance was also performed based on dynamic programming.

4.1   Evaluation with Simulated Image

Fundamental property of LLE was tested in conditions similar to the problem setting. Virtual keypoints are generated as indicated in Fig. 5. It was assumed that an object and the robot hand is captured in a image frame of 400×400 [pix] size. There were 10 keypoints to be detected on the object, 10 on the robot hand and 5 in background. The positions of the object and the hand were varied with uniform distribution for collecting samples. Total number of images was set as $N = 1000$. Number of keypoints was set as $m = 25$. Thus, data matrix for LLE was $H \in \mathbb{R}^{50 \times 1000}$. Dimension of the mapping was set as $d = 3$. To simulate matching error of keypoints, position information of 10 % of the keypoints in the data vector were removed. That is, 10 % of the elements of H was replaced by the average value of positions of the corresponding keypoint.

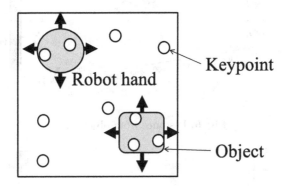

Fig. 5. Simulated keypoints

The result of mapping by LLE is depicted in Figs. 6 and 7. The two figures show the same point information from different perspectives. Y_1, Y_2 and Y_3 in the figures correspond to low-dimensional vector y in (3.6) and hence they do not have units. The colors of the points denote distances between the object and the hand in the corresponding images (Note that original distance information in pixel with maximum 550 pixel was converted to 64 levels.). It can be seen in the figure that one direction in the feature space reflect the distance between the object and the hand.

4.2   Evaluation with Real Image of Humanoid

The evaluation in the previous section did not include keypoints extraction and matching. In the experiment with the humanoid robot, the proposed method described in Sect. 3 was tested. Image size was 640×480. The number of nodes of SOM was set as 6×6. Joint angles ϕ_1 and ϕ_2 were changed an interval of 2 [deg]. Position of the object was changed simultaneously and totally 732 images were taken. (Fig. 8 in original manuscript was removed.) After labeling (Algorithm 1) and integration of labels (Algorithm 2), 1674 labels were obtained.

Images assigned to a node as example are shown in Fig. 8. It can be seen that similar images, corresponding to close positions of the object and the hand, were assigned to a node. Two examples of matching of keypoints are shown in Fig. 9. All of the four images were assigned to a node, while the upper images are the same ones. Lines in the figure indicate matching of keypoints. It can be

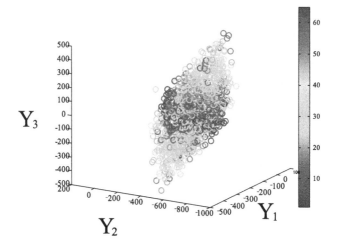

Fig. 6. Distance from object

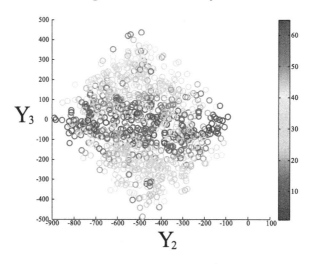

Fig. 7. Distance from object (different perspective)

Table 1. Discrimination of collision with LLE

	Collision [%]	No collision [%]
Recognized as collision	95/115 [82.6]	111/617 [18.0]
Recognized as no collision	20/115 [17.4]	506/617 [82.0]

understood that matching results were different even within images belonging to the same node.

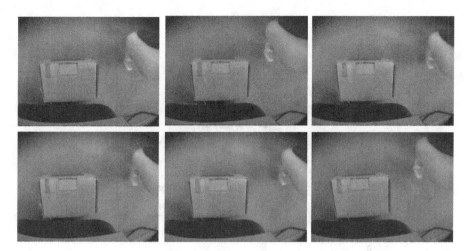

Fig. 8. Images stored in a node in SOM

Table 2. Discrimination of collision with PCA

	Collision [%]	No collision [%]
Recognized as collision	63/115 [54.8]	132/617 [21.4]
Recognized as no collision	52/115 [45.2]	485/617 [78.6]

Fig. 9. Matching result of SIFT keypoints within a node

Figure 10 shows 3-dimensional mapping obtained by the proposed method. Each point (circle or cross) indicates a vector obtained by converting vector x_i by LLE. Cross indicates that the image corresponds to a situation where the hand is contacting with the object. Circle indicates that there is no contact. It can be

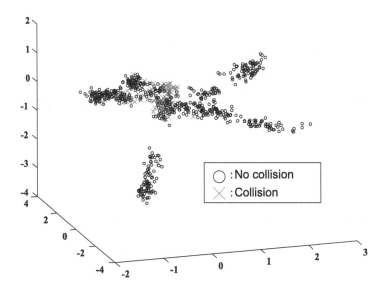

Fig. 10. Obtained mapping with LLE

seen that crosses are concentrating around a certain region. Distance between the object and the hand, however, could not be seen in the obtained map.

Test images, which were not contained in the images for training (generating LLE mapping), were mapped onto the obtained space. Boxes in Fig. 11 indicates test samples, where corresponding images are also displayed. It can be seen that image in which the hand is the most distant from the object is located at the furthest from the region with dense crosses. Images in which the hand is closer are gradually located closer in the mapped space. But there is a jump at the last step to contact with the object into the region with dense crosses.

Using the obtained map, classification of collision was evaluated. Collision of the hand with the object was classified by whether a sample is included in the sphere whose center is the average of the crosses. The optimal radius was set as $r = 0.74$, which was found empirically so that the discrimination performance is the best. Table 1 shows the classification result.

For comparison, a linear mapping was also implemented. Figure 12 shows the result of mapping with PCA (Principal Component Analysis) using the same data matrix. Crosses, corresponding to contact of the hand with the object, are more dispersing compared with Fig. 10. Classification result with PCA is shown in Table 2. It can be seen that consideration of nonlinearity brings conspicuous difference of classification performance.

4.3   Obstacle Avoidance

A sequence of snapshots of motion generated by DP described in Sect. 3.3 is shown in Fig. 13. Grid sizes for the discrete state space was set as $N_{s1} = 8$

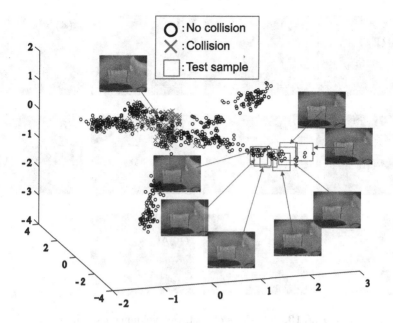

Fig. 11. Prediction of collision with test sample

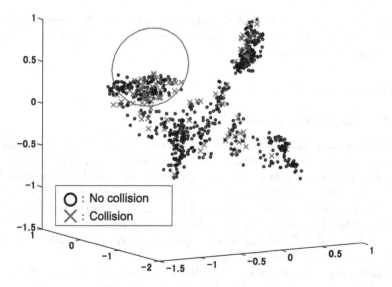

Fig. 12. Mapping result with PCA

and $N_{s2} = 12$. Large negative reward was given for collision based on (3.7), where collision was judged by a correct recognition result for images adopted in Table 1. (1) in the figure denotes the initial configuration of the robot hand. The tip of the hand is located above the object in (11), corresponding to the target

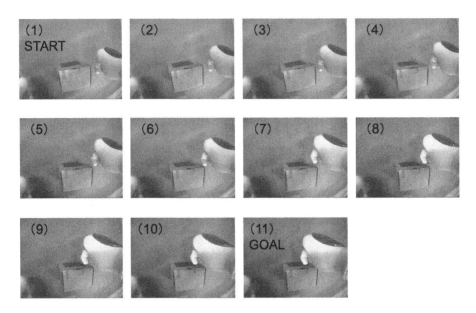

Fig. 13. Snapshots of obstacle avoidance motion

configuration. It can be seen that the robot hand could reach a destination while avoiding collision with the object, given that appropriate evaluation of closeness (or collision) with the object.

5   Discussion

5.1   Labeling of Keypoints

The labels of keypoints obtained by the proposed method was still numerous even after integration of Algorithm 2. It is possible to consider reliability of the keypoints by evaluating frequency of appearance. It should be also considered that there are not so many keypoints stably detected on the hand of NAO. From Fig. 9 it is also clear that mismatching of keypoints was often happening, which deteriorated the LLE mapping. Not only improving reliability of image features (*e.g.*, using PCA-SIFT [8]) but also applying multiple kinds of features will be important to generate good data matrix.

Mismatching of keypoints is substantially inevitable when a part of the robot changes its posture. Therefore, it will be important to expand the framework to a more flexible one, which can continuously map a vector whose elements are partly lost.

5.2   State Space Representation Toward Motion Generation

Motion of the robot hand could be generated by dynamic programming framework, but it is not sufficient as a framework for robot motion generation since

LLE space was not fully utilized for evaluation of relation between the robot hand and the object. In the experiment described above, only image information corresponding to collision was utilized for collision avoidance. However, when LLE reflect more precise relation between the object and the hand, contact of the hand with the object can be predicted using internal simulation using the LLE space.

It is possible to extract keypoints that are relevant to the robot hand by finding correlation between motion of the keypoints and control command (presented in [21,22]). Using the technique, virtual input vector can be generated by changing only position information of keypoints that are relevant to the robot body. The proposed LLE state space can be further utilized by this extension.

6 Conclusions

In this paper, an adaptive approach to building representation of a robot for motion planning, especially collision avoidance, based on a manifold learning method. This approach does not require any specific knowledge on the robot and its environment regarding the collision avoidance motion planning, while only relying on actually-observed sensor information.

In the evaluation of simulated image vectors, it was verified that the distance between the robot hand and the object was reflected in the map. In the evaluation of experiment with real images, the robot could classify images whether the robot is colliding with the object based on the obtained mapping. Moreover, manifold learning turned out to be superior to linear dimensionality reduction, PCA. The proposed framework was demonstrated through motion generation of collision avoidance. Further improvement of prediction of collision and evaluation of closeness to collision will be required for more general and various motion generation. Apart from motion planning of collision avoidance, e.g., manipulating objects, additional structures such as one which enables flexible extention of robot body schema (e.g.,[23]) will be required for approaching to more autonomous robots.

As a next step, it will be required to extend the idea of bottom-up construction of a low-dimensional space to the case where features frequently disappears. Probabilistic framework (e.g. [10]) can be a promising tool for this extension.

Acknowledgment. This work was partly supported by Kayamori Foundation of Informational Science Advancement.

References

1. Aldebaran Robotics. Technical Specifications Document (2009). http://www.aldebaran-robotics.com/
2. Argall, B.D., Chernova, S., Veloso, M., Browning, B.: A survey of robot learning from demonstration. Robot. Auton. Syst. **57**(5), 469–483 (2009)
3. Minato, T., Thomas, D., Yoshikawa, Y., Ishiguro, H.: A model to explain the emergence of imitation development based on predictability preference. IEEE Trans. Autonomous Mental Develop. **4**(1), 17–28 (2012)

4. Theodorou, E., Buchli, J., Schaal, S.: A path integral approach. In: Proceedings of IEEE International Conference on Robotics and Automation, pp. 2397–2403 (2010)
5. Sugimoto, N., Morimoto, J.: Application to humanoid robot motor learning in the real environment. In: Proceedings of IEEE International Conference on Robotics and Automation, pp. 1311–1316 (2013)
6. Minato, T., Asada, M.: Towards selective attention: generating image features by learning a visuo-motor map. Robot. Auton. Syst. **45**(3–4), 211–221 (2006)
7. Prankl, J., Zillich, M., Vincze, M.: 3d piecewise planar object model for robotics manipulation. In: 2011 IEEE International Conference on Robotics and Automation (ICRA), pp. 1784–1790 (2011)
8. Ke, Y., Sukthankar, R.: A more distinctive representation for local image descriptors. In: Computer Vision and Pattern Recognition (2004)
9. Kobayashi, Y., Okamoto, T., Onishi, M.: Generation of obstacle avoidance based on image features and embodiment. Intl. J. Robot. Autom. **24**(4), 364–376 (2012)
10. Somei, T., Kobayashi, Y., Shimizu, A., Kaneko, T.: Clustering of image features based on contact and occlusion among robot body and objects. In: Proceedings of the 2013 IEEE Workshop on Robot Vision (WoRV2013), pp. 203–208 (2013)
11. Kohonen, T.: Self-Organizing Maps. Springer, Heidelberg (1995)
12. Lowe, D.G.: Object recognition from local scale-invariant features. In: Proceedings of IEEE International Conference on Computer Vision, vol. 2, pp. 1150–1157 (1999)
13. Lungarella, M., Metta, G., Pfeifer, R., Sandini, G.: Developmental robotics: a survey. Connect. Sci. **15**, 151–190 (2003)
14. Morimoto, J., Nakanishi, J., Endo, G., Cheng, G., Atkeson, C.G., Zeglin, G.: Poincaré-map-based reinforcement learning for biped walking. In: Proceedings of IEEE International Conference on Robotics and Automation (2005)
15. Oudeyer, P.Y., Kaplan, F., Hafner, V.: Intrinsic motivation systems for autonomous mental development. IEEE Trans. Evol. Comput. **11**(2), 265–286 (2007)
16. Saul, L.K., Roweis, S.T.: Think globally, fit locally: unsupervised learning of low dimensional manifolds. J. Mach. Learn. Res. **4**, 119–155 (2003)
17. Stoytchev, A.: Some basic principles of developmental robotics. IEEE Trans. Autonomous Mental Develop. **1**(2), 122–130 (2009)
18. Sutton, R.S., Barto, A.G.: Reinforcement learning: an introduction (Adaptive Computation and Machine Learning). In: A Bradford Book (1998)
19. Kober, J., Bagnell, D., Peters, J.: Reinforcement learning in robotics: a survey. Intl. J. Robot. Res. **11**, 1238–1274 (2013)
20. Weng, J., McClelland, J., Pentland, A., Sporns, O., Stockman, I., Sur, M., Thelen, E.: Autonomous mental development by robots and animals. Science **291**, 599–600 (2001)
21. Fitzpatrick, P., Metta, G., Natalc, L., Rao, S., Sandini, G.: Learning about objects through action - initial steps towards artificial cognition. In: Proceedings of IEEE International Conference on Robotics and Automation, pp. 3140–3145 (2003)
22. Stoytchev, A.: Toward video-guided robot behaviors. In: Proceedings of the 7th International Conference on Epigenetic Robotics, pp. 165–172 (2007)
23. Kobayashi, Y., Hosoe, S.: Planning-space shift motion generation: variable-space motion planning toward flexible extension of body schema. J. Intell. Robot. Syst. **62**(3), 467–500 (2011)

The Existence of Two Variant Processes in Human Declarative Memory: Evidence Using Machine Learning Classification Techniques in Retrieval Tasks

Alex Frid[1(✉)], Hananel Hazan[2], Ester Koilis[1], Larry M. Manevitz[1], Maayan Merhav[3], and Gal Star[1]

[1] Computer Science Department, University of Haifa, Haifa, Israel
alex.frid@gmail.com, esterkoilis@yahoo.com,
manevitz@cs.haifa.ac.il, gal.star3051@gmail.com
[2] Network Biology Research Laboratory, Technion, Haifa, Israel
hananel@hazan.org.il
[3] German Center for Neurodegenerative Diseases (DZNE),
Magdeburg, Germany
themaayan@yahoo.com

Abstract. This work use supervised machine learning methods on fMRI brain scans, taken/measured during a memory-*retrieval* task, to support establishing the existence of two distinct systems for human declarative memory ("Explicit Encoding" (EE) and "Fast Mapping" (FM)). The importance of using retrieval is that it allows a direct comparison between exemplars designed to use EE and those designed to use FM. This is not directly available under *acquisition* tasks because of the nature of the purported memory systems since the tasks are necessarily somewhat distinct between the two systems under acquisition. This means that there could be a confounding of the distinction in the *task* with the difference in the representation and mechanism of the internal memory system during analysis. *Retrieval* tasks, on the other hand allow for identity of task. Thus this work fills a lacuna in earlier work which used memory acquisition tasks. In addition, since the data used in this work was gathered over a two day period, the classification methods is also able to identify a distinction in the *consolidation* of the memories in the two systems. The results presented here clearly support the existence of the two distinct memory systems.

Keywords: Machine learning · Classification · Functional Magnetic Resonance Imaging (fMRI) · Feature selection · Support vector machines · Decision trees · Radial basis function kernel · Declarative memory · Consolidation · Semantic memory · Informational biomarkers

1 Introduction

Human declarative memory is defined as the conscious recollection of facts and events [1]. Under the theory of declarative memory systems, novel information is encoded into the memory using, amongst other brain parts, the hippocampus [2]. In this study,

© Springer-Verlag Berlin Heidelberg 2016
N.T. Nguyen et al. (Eds.): TCCI XXIV, LNCS 9770, pp. 117–133, 2016.
DOI: 10.1007/978-3-662-53525-7_7

standard, hippocampal dependent memory is represented by "Explicit Encoding (EE)" procedure. According to memory transformation theories of declarative memory, the encoded information is slowly transferred from the hippocampus to the neo-cortex where it becomes permanently stored [3, 4]. Over time, the initially hippocampal dependent memories become independent of the hippocampus. It has been suggested that this re-organization process is done during sleep [5].

Amongst toddlers, the process of rapid language acquisition occurs prior to the full development of the hippocampus [6, 7]. Moreover, some evidence from hippocampal injured subjects demonstrated an ability to acquire information which seems to have declarative-like characteristics, despite severe damages in the hippocampus [8, 9] and so, must involve a different brain network than the one which engages the hippocampus This alternative learning mechanism can be acquired via "Fast Mapping (FM)". It is unclear if the memory representations following FM undergo consolidation processes overtime, as do memories gained through EE. However, since it was shown that patients with hippocampal damages as well as healthy controls could learn and store information acquired via FM [8, 9], the scheme used to explain memory consolidation of declarative memories cannot be applied for FM in a straightforward manner.

In this work we aim to demonstrate the distinctiveness of brain systems, which support EE and FM memory process, by extracting activity patterns directly from brain data, using Functional Magnetic Resonance Imaging (fMRI) method. fMRI captures information from thousands of different localities (voxels) of the brain, simultaneously. Then multivariate pattern analysis approach (MVPA) [10] utilizes these activities by looking for changes in Blood Oxygen Level-Dependent (BOLD) signal across different voxels. Different methods can be used for analysis on such complex data depending on the question of study (retrieval or decoding stimuli, mental states, behavior and other variables of interest). A growing number of studies [11–14] shows some of the capability in using machine learning methods for analysis of neuroimaging data. Moreover, the feasibility to achieve successful results using machine learning on fMRI multivariate data is not trivial and relies on the sensitive choice of features to be considered in the analysis.

In this work we focused on classification techniques, in particular using SVM and decision trees. The overriding motif was that if machine learning can distinguish between tasks from the fMRI data, then they are performed differently.

2   Related Work

The mechanism of FM was examined among healthy individuals [14, 15]. It was shown that two learning mechanisms, EE and FM, can be discriminated from fMRI data during memory acquisition, using machine learning based classifiers. In addition, scans taken while memory acquisition were tested for success in a consecutive retrieval task, outside the fMRI machine. Successful accuracy results were achieved when identifying scans corresponding to correct and incorrect retrieval, within the EE group and within the FM group, for each participant separately and cross-participant.

However, the different nature of the procedures used for acquisition of information (EE and FM), did not allow for complete control over the task with regard to the behavioral experience. Therefore, the possibility remained that the successful classification obtained in the experiment was a result of differences in the acquisition procedures and not in the learning mechanisms.

To overcome this limitation, in this work we examine data obtained in another study [16]. There, the neural correlates of FM and EE were explored during a retrieval procedure, designed to be identical for both mechanisms. In addition, the study focused on overnight re-organization of memory representations, following both EE and FM, by comparing recent memories to remote ones (obtained in the previous day). Findings suggested that, despite the identical retrieval tasks, memories that were gained through FM induced distinct neural substrates from those involved EE [16]. While retrieval of data learned through EE engaged the expected hippocampal and vmPFC related network, retrieval of information acquired through FM immediately engaged an ATL related network, typically supporting well-established semantic knowledge. In addition, analysis of neuroimaging data associated with EE showed the expected overnight changes in network connectivity where for FM minimal overnight changes were presented. The analysis was performed by a multivariate technique of Spatiotemporal Partial Least Squares (PLS), helping to identify assemblies of brain regions that co-vary together.

3 Current Study

In this study, fMRI brain data were captured during the retrieval of memories, acquired through either EE or FM. The goal is to provide a biomarker directly from these fMRI scans using machine learning methods. Such classification ability based on the neural activity data gives strong evidence for the existence of distinct neural processes associated with EE and FM.

Multivariate classification is performed on fMRI features obtained during memory retrieval where tasks performed by the participants are identical for EE and FM. We also perform classification to explore the overnight re-organization processes following both learning mechanisms. Classification was performed over brain scans which were acquired either 30 min before scanning (recent memory) or a day before scanning (remote memory).

Regarding the distinction between the two memory processes during retrieval, we address two questions:

- Is it possible to distinguish between the two learning modes (i.e. EE and FM) based on neural activity information, collected during the retrieval of memories?
- Is it possible to distinguish between items learned recently and remotely?

4 Experiment Procedure

4.1 Participants

The experiment, full details of which can be seen in Merhav et al. [16], was conducted in Rotman Research Institute at Baycrest, Canada. Here, we mention the salient points.

22 participants were recruited and randomly assigned to one of the two groups (EE or FM). All participants were English native speakers, right-handed and had no history of neurological or psychiatric disorders and no learning disabilities. A written informed consent was obtained according to Baycrest's Research Ethics Board's guidelines. Gender and age distributions (10 females in each group) were similar in the FM and in the EE groups, respectively. The two groups also did not differ on the number of years of education, I.Q. estimates and WMS-III Verbal Paired Associates retention.

4.2   Experiment Paradigm and Procedure

22 healthy adult participants were randomly assigned of one to two groups (EE or FM). On day 1 the participants learned 50 new unfamiliar picture-word associations. On day

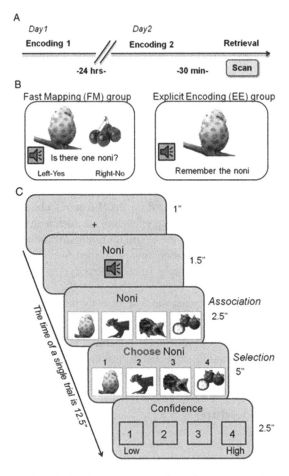

Fig. 1. (A) The experiment structure. (B) Examples of acquisition through FM (left) and through EE (right). (C) Retrieval test design which took place inside the fMRI scanner.

2 (24 h later) they learned another set of 50 new picture-word associations. A retrieval memory test for all the 100 new picture-word associations took place 30 min after the acquisition of the second set of associations. During the retrieval, brain activity was scanned (Fig. 1A). Therefore, the participants were tested on both recently and remotely encoded information. The two learning tasks (EE / FM) were designed differently due to different nature of both learning procedures (Fig. 1B).

The retrieval task was designed as an event related fMRI experiment in which memory for all 100 items was assessed via an associative four-alternative forced choice recognition task. The retrieval procedure was identical for EE and FM as it was performed inside the scanner (Fig. 1C). Retrieval trial of each item was 12.5 s long and contained the following intervals: blank screen (1 s), target label as text and auditory input (1.5 s), 4 choice pictures appeared on screen, below the target label (2.5 s), the word "choose" appeared onscreen and participants had to respond by selecting the appropriate key (5 s), confidence rating (2.5 s).

The experiment was intentionally designed to have participants perform either EE or FM, rather than perform both EE and FM tasks. It was important that learning through FM will be implicit and unintentional, so participants should not know that the task involves memory. However, in EE, participants are explicitly asked to remember the name of the item.

4.3   Data Acquisition and Pre-processing

The participants were scanned using the Siemens Trio 3T scanner, at Baycrest Institute. They acquired T2*-weighted images, covering the whole brain using an echo-planar imaging (EPI) sequence of 50 slices, with repetition time (TR) of 2500 ms, echo time

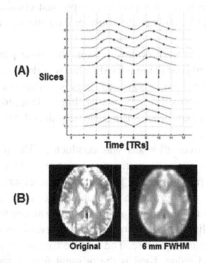

Fig. 2. Examples for pre-processing steps on fMRI data. (A) Correction of individual's hemodynamic responses slices acquired aligned to the exact same time [18]. (B) Performance of spatial smoothing on fMRI volume taken from single participant.

(TE) of 27 ms, 64 × 64 matrix, slice thickness of 3.5 mm and a field of view (FOV) of 200 mm. The procedure was designed as an event related fMRI study.

The pre-processing steps included conversion to 4-dimensional AFNI format [25], followed by slice timing correction using the first slice as a reference (Fig. 2A), latter movement correction for unintended head motions and spatial smoothing with 6 mm FWHM Gaussian kernel to increase signal-to-noise ratio was applied (Fig. 2B). Finally, the individual participant's data was converted to a standard coordinate system (Talairach [26]) to allow data analysis across individuals.

The scanning of each participant was done during four runs, creating a joint dataset out of four time-series datasets, with approximately 150 data volumes each of size 109 × 91 × 91, resulted as a dataset with approximately 600 data volumes. Therefore, each data volume (data point) contained 1490580 different voxels. We demonstrate the structure of the collected data in Fig. 3.

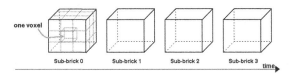

Fig. 3. 4-dimensional structure of AFNI format BRIK (Cox, 1996) file including 3-dimensional dataset over time sequence.

5   Methods

The data points used for analysis were constructed using scan data obtained for TR=2. This temporal cut was selected after performing pre-test classification as suggested in Atir-Sharon et al. work [14], taking into consideration the accordance to the expected HRF response.

We performed further pre-processing over the time-series data. At first, all non-brain voxels were removed using a mask. This was done by selecting voxels from the fMRI dataset that correspond to non-zero elements in the mask (creating data points of approximately 200,000 voxels). Afterwards, linear de-trending was performed on each participant's data set and for each run separately in order to remove low frequency signal intensity drifts.

Then, normalization over all scans was conducted. The normalization was done voxel-wise using z-score for each participant separately. In our case, the combined dataset involved scans from different groups and participants taken from different distributions. Therefore, transformation of features from different scales to a single scale, with consideration to the original distributions, was needed. The z-score method considers the different distribution characteristics of every group [17], hence, it was chosen as the normalization procedure. The z-score formula is presented in (1), where z-val is the new z-scored value, f-val is the original feature value and (μ, σ) are the mean and standard deviation values:

$$z\text{-val} = (f\text{-val} - \mu)/\sigma \tag{1}$$

For the mean and standard deviation computation in the z-score equation, several assignments were tested: (i) from all scans in the dataset; (ii) from individual participants' scans and (iii) from the distribution of scans marked as control (baseline) in the training set. Best classification results were achieved by using the mean and standard deviation normalization as computed from the distribution of baseline scans (option (iii)).

Each volume was represented as an individual data point in the dataset (i.e. each voxel was considered as a feature). Since the amount of scans from EE and FM groups was not equal, counter-balancing of the dataset was performed. This was done by randomly sampling data points from the smaller group. This method was applied only on the training set. (Otherwise, more weight would have been given to prediction accuracies of duplicated data points against weight of accuracies for data points that were not duplicated.) Therefore, the testing set was left untouched.

Machine learning classification techniques were used for data analysis. Considering the high dimensionality of data used in the current study, feature selection procedures were performed. The purpose of these procedures is to reduce the number of feature-voxels used for multivariate classification analysis. Such reduction is meant to remove irrelevant voxels and to improve training time.

There are several generic methods for selecting informative features. We aimed to select the features that best discriminate between conditions based on their activation values. It was achieved by ranking the importance of each feature according to the ANOVA F-score value obtained for the corresponding contrast (e.g., Correct vs. Incorrect Retrieval in FM condition) comparisons.

To find the optimal subset of features for analysis, we examined different sizes of features sets starting from 10 features to full brain scans. Eventually, based on the obtained accuracy values, the top 1000 voxels with highest F-scores were selected. This relatively large number of features was chosen to include some weakly infor-mative voxels which can contribute to an increase in classification rates [19].

In Fig. 4, we illustrate the extracted features in the form of a brain map. In this example, we display selected subset of features for retrieval (Correct vs. Incorrect) classification. This brain map is an example showing the voxels selected on a specific individual's fMRI data that belongs to the FM group. Although not reproduced here, a detailed list of the top ANOVA chosen features for FM task shows the ATL area well represented and this accords with current ideas of the location of the FM activity.

In the first stage, a cross-validation classification scheme using Support Vector Machine classifier [20] with RBF (Radial Basis Function) kernel [21] was applied to the selected features. Parameters that are not learnt directly within estimators can be set by searching a parameter space for the best cross-validation score. Grid search for C and gamma parameters was performed in the ranges of 2^{-5} to 2^{15} and 2^{-15} to 2^{3} respectively. Grid search was executed before training on a training portion of the dataset to achieve increase in accuracy rates. A pseudo-code for the performed grid search is presented in Fig. 5. In all runs parameters C and gamma were set to 1 and 2^{-3} respectively.

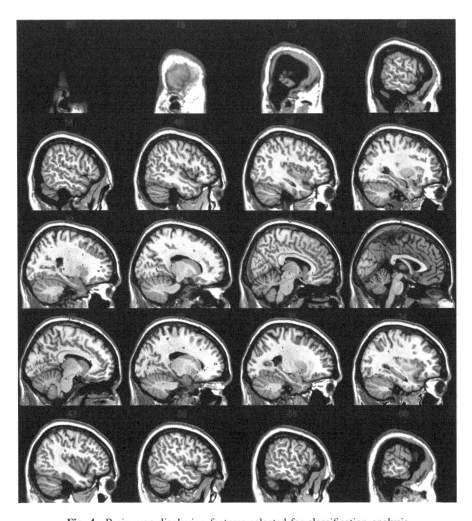

Fig. 4. Brain map displaying features selected for classification analysis

```
for c in [2⁻⁵, 2⁻³,...,2¹⁵]:
  for g in [2⁻¹⁵, 2⁻¹³, ..., 2⁵]:
   for train, test in partition:
     model = svm_train(train, c, g)
     score = svm_predict(test, model)
     cv_list.insert (score)
    scores_list.insert(mean(cv_list),c,g)
print max(scores_list)
```

Fig. 5. Pseudo-code for the grid search procedure.

In cases where the testing set consisted of scans that were taken from one group only (i.e. all scans were EE or all scans were FM), a decision making function was applied. We used majority voting method as a decision making function, defined as follows: if the majority of the scans were rated correctly per participant, the accuracy was set to 1, otherwise, the accuracy was set to 0.

The software used for the classification was developed using Python programming language and based on LibSVM [22] and PyMVPA software packages [23]. In Fig. 6 we present a complete analysis flow diagram including all the relevant pre-processing and processing stages.

Later we saw that this approach has several disadvantages in the context of this problem: (i) a best kernel for the specific problem needs to be found, (ii) an exhaustive parameter search need to be applied in order to optimize the parameters of the kernel for a specific problem, and (iii) the initial dimension of the data is already high (1000 features after the feature selection process), so projection of this data into a much higher space results in sparsity (due to the relatively small set of the data points) and hence a poorer generalization during the margins maximization process (which in turn results in a relatively high standard deviation). (In addition, as a practical matter, the computational resources for SVM are rather high which limits the freedom of experimentation of variants. In our work, the SVM methodology including the grid search for the parameters "C" and "Gamma" as well as the cross-validation means the time was of the order of several hours for each experiment. Further, it is important to notice

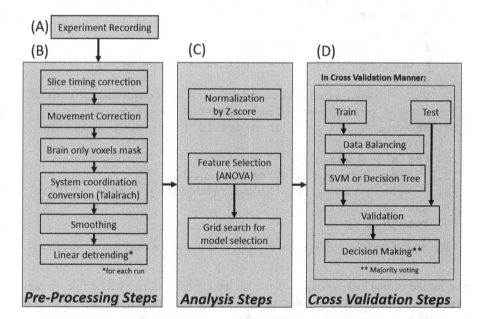

Fig. 6. Schematic diagram of the steps performed for whole brain analysis procedure. It consists of the following stages: (A) The initial stage representing the neuroimaging data delivery. (B) The pre-processing stage. (C) Data reduction stage: reducing data variability efficiently by feature selection. (D) Learning stage: performing multiple times by cross validation procedure.

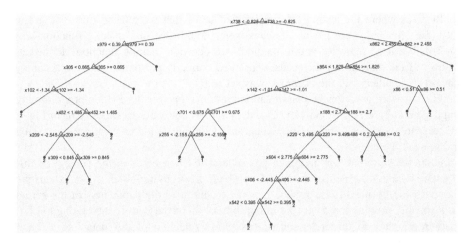

Fig. 7. An example of a Decision Tree trained on EE versus FM, on 70 % of the data and limited to minimum of 7 observations per tree leaf, the result for this tree on the test data is 70 %.

```
generateTree(scans)
Input: scans - fMRI scans (containing 1000 best voxels)
1.  Tree={}
2.  If scans is "pure" OR stopping criteria met then
3.     Terminate
4.  end if
5.  for all voxels ∈ scans do
6.     compute Gini's index criteria if we split on 'v'
7.  end for
8.  vbest = best voxel according to the computed criteria
9.  Tree = create decision node that tests vbest
10. Scanssub = induced sub-datasets from scans based on
         vbest
11. for all scanssub do
12.    Treesub = generateTree(scanssub)
13.    Attach Treesub to the corresponding branch of Tree
14. end for
15. return Tree
```

Fig. 8. The Decision tree algorithm that used in this study for fMRI data classification

that for the current task (giving evidence for the existence of two declarative memory systems) it is not a priority to optimize the classification capability of the system. That is, it suffices to show that the systems can be separated in a significant manner.

Taking into account all of these issues we decided to also use a version of Decision trees [27] with a Gini index [28] as a splitting criteria, a competitive machine learning tool. That is, in our case, the splitting value for each node is calculated by choosing the value for each potential feature that maximizes the homogeneity of the "Gini impurity function" over the split. (See [28] for a full description.) Note that machine learning generation of decision trees have a long history [27, 28]. Its advantages include (i) it is typically created in a "greedy" fashion and so needs much less computational resources, both time and memory (ii) it creates its tree in the original feature space and so this helps in later understanding of the classification. In addition, the use of alternative methods gives some additional insight into the results. The pseudo-code for tree creation is given below in Fig. 8. (While the time complexity of such a tree can be as the square of the number of features; in actual practice we found that the depth of the tree is quite small, and so it scales linearly as the number of features. See Fig. 7 for a typical tree. Including the cross validation calculations results in calculations that takes less than a minute. We mention that this speedup suggests the possibility of using a "wrapper approach" for the choice of features instead of the ANOVA and in future work, we will explore this possibility.)

6 Results

6.1 Memory Performance

In the information retrieval test as fully presented in Merhav et al. [16], correct response rates for the recent and for the remote associations were significantly above chance (binomial tests, $p < 0.0001$, for both times-of-acquisition, in both learning groups). Overall, participants from the FM group were less successful in retrieval, compared to those from the EE group, in both the recent and the remote conditions (F $(1,30) = 12.2$, $p < 0.005$).

In both groups, recent items were better recognized than remotely presented items $(F(1) = 9.12$, $p = 0.005)$ with no significant interaction between the time of acquisition and the learning mode $(F(1,15) = 0.334$, $p = 0.565)$.

6.2 Classification

First, we addressed the question of classifying scans obtained during correct and incorrect retrieval. Using the proposed classification scheme, we performed 4-fold (leave one run out) cross-validation within participants. However, the mean values of classification accuracy were close to the chance level for both groups (EE and FM). We theorized that the reason could be the existence of two additional different sub-groups, recent and remote acquisition, within each of the initial groups.

Accordingly, we classified correct and incorrect scans within each possibility: EE recent, EE remote, FM recent, FM remote. For each possibility we chose 10 % of all data points randomly as a testing set. The rest of the data points were used for training. Then, 10-fold cross validation was performed. We report the values for mean and standard deviation of classification accuracy over 10 cross-validation folds for EE in Table 1 and for FM in Table 2.

These results show that a trained classifier was able to distinguish scans obtained during correct and incorrect word retrieval within each group. The accuracy is higher for classification of scans for words learned recently, rather than for words learned remotely. Furthermore, the discriminating ability is better within EE group rather than within FM group. From Tables 1 and 2 we see that some significant change has taken place in the activation overnight for EE; and not much can be seen in FM.

This is in accordance with the current idea of how EE and FM are stored; i.e. FM is directly stored in cortex and EE initially involves the hippocampus and over some time undergoes consolidation into the cortex. It is possible that there is more variation in the specific activations in the prior to consolidation voxels than in the cortical ones; accordingly more of the ANOVA voxels are in the hippocampus. This would help account for the lowered accuracy in EE after time as seen in Table 1 and the continued lower accuracy in FM in Table 2. However, clarification of this point will require further experimentation.

Next, we classified whether the process used for information acquisition was EE or FM using only scans from the successful retrieval attempts in the behavioral experiment. We chose randomly 10 % from all these scans of all participants as a testing set. The rest of these scans were used as a training set. Under this protocol there is substantial training data from each participant. Table 3 shows that in this case EE and FM scans can be very well distinguished.

Table 1. Correct vs. Incorrect classification within Explicit Encoding (EE) using 10-fold cross validation.

	Mean accuracy	Standard deviation
Recent	0.708	0.09
Remote	0.584	0.067

Table 2. Correct vs. Incorrect classification within Fast Mapping (FM) using 10-fold cross validation.

	Mean Accuracy	Standard deviation
Recent	0.599	0.063
Remote	0.55	0.068

Table 3. EE vs. FM (using only scans with correct retrieval) random scan selection cross-validation.

Testing set selection method	Mean accuracy	Standard deviation
Random selection	0.937	0.069

These results raise the question of whether the representation of all the participants in the training set is crucial to the classification success. That is, can a machine learning classifier, trained over the collected data, successfully distinguish which label to assign to a new individual's scan, despite the fact that the classifier has never seen data from this participant. To answer this question, we performed a leave-one-participant-out classification. This was done across all 16 (one from EE and one from FM) participants in a cross-validation manner (leave one out). Note that per iteration, the scans in the testing set are all EE or all FM. Therefore, we were able to use the majority voting method for this analysis. The results averaged across all participants presented in Table 4.

Table 4. EE vs. FM (using only scans with correct retrieval scans) across participants using 16-fold cross-validation

Testing set selection method	Mean accuracy	Standard deviation
Leave one participant out	0.638	0.07

Table 5. EE versus FM Confusion Matrix, average accuracy is 73 %

	EE	FM
EE	67.88 ± 4.68	32.12 ± 4.68
FM	22.68 ± 3.65	77.32 ± 3.65

Table 6. Recent (EE versus FM) Confusion Matrix, average accuracy is 70 %

	EE	FM
EE	0.6425 ± 0.0738	0.3575 ± 0.0738
FM	0.2516 ± 0.0568	0.7484 ± 0.0568

Table 7. Remote (EE versus FM) Confusion Matrix, average accuracy is 68 %

	EE	FM
EE	62.56 ± 7.54	37.44 ± 7.54
FM	25.71 ± 5.15	74.29 ± 5.15

Table 8. Recent versus Remote Confusion Matrix, average accuracy is 61 %

	Recent	Remote
Recent	60.45 ± 4.29	39.55 ± 4.29
Remote	38.99 ± 4.97	61.01 ± 4.97

Table 9. EE (Recent versus Remote) Confusion Matrix, average accuracy is 63 %

	Recent	Remote
Recent	61.83 ± 6.28	38.17 ± 6.28
Remote	36.81 ± 6.22	63.19 ± 6.22

Table 10. FM (Recent versus Remote) Confusion Matrix, average accuracy is 50 %

	Recent	Remote
Recent	49.81 ± 6.90	50.19 ± 6.90
Remote	49.11 ± 6.64	50.89 ± 6.64

Tables 5, 6, 7, 8, 9 and 10 presents the classification results using 1000 best features selected by ANOVA F-score from the whole brain using the alternative Decision Tree methodology. To avoid over-fitting a minimum of 7 observations per tree leaf was required. All the results were generated using 80 cross-validation cycles with randomly chosen scans for testing and training. (The division was 30 % for testing and 70 % for training.)

Tables 5, 6 and 7 show how the decision trees succeed in separating EE from FM. In all cases (both all data and separated by time since acquisition) they can be separated in a significant fashion.

Tables 8, 9 and 10 relate to the issue of consolidation. The general conception ("standard model" [1] is that EE undergoes a transition between storage very dependent on the hippocampus to one based on the cortex; whereas under FM this process may be quite different. Comparing Tables 9 and 10, this is borne out.

In summary, we see that these results strongly affirm the distinction between EE and FM. In addition, we see that the retrievals between recent and remote are distinguishable only in EE. This indicates that re-organization takes place in the time frame of the experiment for EE examples; while we were unable to discern such a distinction for the FM mechanism. This corresponds to the theoretical expectation of the two system and further supports their existence.

7   Discussion and Conclusions

In this work, we showed that it is possible to identify correct and incorrect retrieval of memories acquired through two learning mechanisms: either Explicit Encoding (EE) or Fast Mapping (FM) directly from neuroimaging data, using machine learning techniques. The findings suggest (Tables 1 and 9) that it is easier to identify retrieval success and failure for information acquired recently rather than for information after a period of time through EE mechanism. At the same time, no significant change (Tables 2 and 10) between retrieval results of recent and remote acquisition was seen within the FM mechanism. This may indicate that FM does not engage reorganization processes during the 24 h since encoding.

It was also observed that one could directly classify which memory system was used regardless of when the memory was acquired (Tables 5, 6 and 7).

Accordingly, the current results provide additional evidence for the existence of two memory formation processes by successfully classifying scans of correct retrievals following EE and FM. Note that the classification results for scans taken from an individual's data, which were not used previously for training, were still significant. These findings suggest that associative learning through FM employs alternative neural pathways to acquire and maintain declarative knowledge. This also indicates that the FM process is eligible for therapeutic approach for people with hippocampal brain injuries.

8 Future Work

Future work should include mapping of the brain regions and extraction of functional networks associated with all four group combinations, EE recent, EE remote, FM recent and FM remote. A list of possible implementation approaches includes constructing brain maps using "searchlight" techniques [12].

Another novel approach is to consider the actual voxels used in the Decision Trees. Looking at a typical example (Fig. 7) shows many interesting aspects. Note that the decision is made by a very small number of voxels. Furthermore, the interaction between the voxels on each path through the tree is clearly indicated. This means that, taking into an account of the location of each voxel used, a careful analysis should indicate the interaction between areas of the brain for each memory system.

In addition, future work should include brain regions correlations tests during the retrieval of memory through EE and through FM in recent and in remote modes. Those correlations would provide information regarding the involvement of the hippocampus and vmPFC regions in the consolidation processes. To achieve that, one may use causality analysis techniques [24] to reveal the causality influences the brain regions, which are involved with each learning procedure, have on each other. This could help reveal new information regarding the mechanism involved in memory consolidation processes of FM.

Acknowlegments. Part of this work appears in the M.Sc thesis of Ms. Gal Star at University of Haifa under the supervision of Prof. Larry Manevitz at the Neuro-Computation Laboratory at Caesarea Rothschild Institute (CRI), Haifa, Israel.

The research is based on data gathered by Rotman Research Institute at Baycrest, Toronto, Canada. The examining of this data was suggested by Dr. A. Gilboa and complements the work of Merhav, Karni and Gilboa [16]. The computational analysis of the data was performed at the Neuro-Computation Laboratory at the Caesarea Rothschild Institute at the University of Haifa, Israel under the supervision of Prof. Larry Manevitz. The authors are listed in alphabetical order.

References

1. Squire, L.R.: Declarative and non-declarative memory: multiple brain systems supporting learning and memory. J. Cogn. Neurosci. **4**(3), 232–243 (1992)

2. McClelland, L., McNaughton, B.L., O'Reilly, R.C.: Why there are complementary learning system in the hippocampus and neo-cortex: insights from the successes and failure of connectionist models of learning and memory. Psychol. Rev. **102**(3), 419–457 (1995)

3. Squire, L.R., Alvarez, P.: Retrograde amnesia and memory consolidation: a neurobiological perspective. Current Opin. Neurobiol. **5**(2), 169–177 (1995)

4. Frankland, P.W., Bontempi, B.: The organization of recent and remote memories. Nature Rev. Neurosci. **6**, 119–130 (2005)

5. Gais, S., Albouy, G., Boly, M., Dang-Vu, T.T., Darsaud, A., Desseilles, M., Rauchs, G., Schabus, M., Sterpenich, V., Vandewalle, G., Maquet, P., Peigneux, P.: Sleep transforms the cerebral trace of declarative memories. Proc. Nat. Acad. Sci. USA **104**(47), 18778–18783 (2007)

6. Bauer, P.J.: Toward a neuro-developmental account of the development of declarative memory. Dev. Psychobiol. **50**(1), 19–31 (2008)

7. Uematsu, A., Matsui, M., Tanaka, C., Takahashi, T., Noguchi, K., Suzuki, M., Nishijo, H.: Developmental trajectories of amygdale and hippocampus from infancy to early adulthood in healthy individuals. PLoS ONE **7**(10), e46970 (2012)

8. Sharon, T., Moscovitch, M., Gilboa, A.: Rapid neocortical acquisition of long-tem arbitrary associations independent of the hippocampus. Proc. Nat. Acad. Sci. USA **108**(3), 1146–1151 (2011)

9. Merhav, M., Karni, A., Gilboa, A.: Neocortical catastrophic interference in healthy and amnesic adults: A paradoxical matter of time. Hippocampus **24**(12), 1653–1662 (2014)

10. Norman, K.A., Polyn, S.M., Detre, G.J., Haxby, J.V.: Beyond mind-reading: multi-voxel pattern analysis of fMRI data. Trends Cogn. Sci. **10**(9), 424–430 (2006)

11. Mitchell, T., Shinkareva, S., Carlson, A., Chang, K.M., Malave, V.L., Mason, R., Just, M.A.: Predicting human brain activity associated with the meanings of nouns. Science **320** (5880), 1191–1195 (2008)

12. Kriegeskorte, N., Goebel, R., Bandettini, P.: Information-based functional brain mapping. Proc. Nat. Acad. Sci. USA **103**(10), 3863–3868 (2006)

13. Nawa, N.E., Ando, H.: Classification of self-driven mental tasks from whole-brain activity patterns. PLoS ONE **9**(5), e97296 (2014)

14. Atir-Sharon, T., Gilboa, A., Hazan, H., Koilis, E., Manevitz, L.M.: Decoding the formation of new semantics: MVPA investigation of rapid neocortical plasticity during associative encoding through Fast Mapping. Neural Plast. **2015**, 17 (2015)

15. Gilboa, A., Hazan, H., Koilis, E., Manevitz, L., Sharon, T.: Two memory systems: identifying human memory encoding mechanisms from psychological fMRI data via machine learning techniques. In: Proceedings of the International Joint Conference on Neural Networks (IJCNN), p. 54 (2011)

16. Merhav, M., Karni, A., Gilboa, A.: Not all declarative memories are created equal: fast mapping as a direct route to cortical declarative representations. Neuroimage **117**, 80–92 (2015)

17. Wiesen, J.P.: Benefits, Drawbacks, and Pitfalls of z-Score Weighting. In: 30th Annual IPMAAC Conference (2006). http://annex.ipacweb.org/library/conf/06/wiesen.pdf, 27 Jun 2006

18. Sladky, R., Friston, K.J., Tröstl, J., Cunnington, R., Moser, E., Windischberger, C.: Slice-timing effects and their correction in functional MRI. Neuroimage **58**(2), 588–594 (2011)

19. Gonzalez-Castillo, J., Saad, Z.S., Handwerker, D.A., Inati, S.J., Brenowitz, N., Bandettini, P.A.: Whole-brain, time-locked activation with simple tasks revealed using massive averaging and model-free analysis. Proc. Nat. Acad. Sci. **109**(14), 5487–5492 (2012)

20. Vapnik, V.: Statistical learning theory. Wiley, New York (1998)

21. Vert, J.P., Tsuda, K., Schölkopf, B.: A primer on kernel methods. Kernel Methods in Computational Biology (2004)
22. Chang, C.C., Lin, C.J.: LIBSVM: a library for support vector machines. ACM Trans. Intell. Syst. Technol. (2011). http://www.csie.ntu.edu.tw/~cjlin/libsvm
23. Hanke, M., Sederberg, P.B., Hanson, S.J., Haxby, J.V., Pollmann, S.: PyMVPA: A python toolbox for multivariate pattern analysis of fMRI data. Neuroinformatics 7(1), 37–53 (2009)
24. Hu, S., Liang, H.: Causality analysis of neural connectivity: New tool and limitations of spectral granger causality. Neurocomputing 76(1), 44–47 (2012)
25. Cox, C.: AFNI: software for analysis and visualization of functional magnetic resonance images. Comput. Biomed. Res. 29, 126–173 (1996)
26. Talairach, J., Tournoux, P.: Co-planar stereotaxic atlas of the human brain. 3-Dimensional proportional system: an approach to cerebral imaging (1988)
27. Breiman, L., Friedman, J., Olshen, R., Stone, C.: Classification and Regression Trees. CRC Press, Boca Raton (1984)
28. Gelfand, S.B., Ravishankar, C.S., Delp, E.J.: An iterative growing and pruning algorithm for classification tree design. IEEE Trans. Pattern Anal. Mach. Intell. 13(2), 163–174 (1991)

Divide and Conquer Ensemble Method for Time Series Forecasting

Jan Kostrzewa[1,2], Giovanni Mazzocco[1,2(✉)], and Dariusz Plewczynski[2]

[1] Institute of Computer Science, Polish Academy of Sciences,
ul. Jana Kazimierza 5, 01-248 Warsaw, Poland
`g.mazzocco@cent.uw.edu.pl`
[2] Centrum of New Technologies, University of Warsaw,
ul.Banacha 2c, 02-097 Warsaw, Poland

Abstract. Time series forecasting have attracted a great deal of attention from various research communities. There are many methods which divide time series into subseries. Information granules, fuzzy clustering and data segmentation are among the most popular methods in this field. However all these methods are designed to recognize dependencies between adjacent points. In order to do so, they divide the time series into time intervals. This imply some limitations in findings strongly non-local dependencies between points scatter across whole time series. The Divide and Conquer ensemble algorithm here presented was designed to address such limitations. The model samples the series into many subseries, searches for possible patterns and finally chooses the most significant subseries for further investigation. Since the prediction error evaluated on the subseries is lower than the one evaluated on the original time-series, the proposed strategy can significantly mitigate the overall prediction error. In order to evaluate the efficiency of our approach we performed the analysis on various artificial datasets. In a real world example our algorithm showed a 3-fold improvement of the accuracy with respect to other state-of-the-art methods. Although the algorithm was designed for time-series forecasting, it can be easily used for noise filtering purposes. Simulations reported in the present work illustrate the potential of the method in this field of application.

Keywords: Time series · Forecasting · Prediction · Data mining · Divide and conquer · Ensemble

1 Introduction

Time series forecasting is a rich and dynamically growing scientific field and its methods are applied in numerous areas such as medicine, genomics, economics, finance, engineering and many others (Wu 2005, Zhang 2007, Zhang 2003, Tong 1983). The Divide and conquer (D&C) approach is a well known paradigm used in algorithm design. It acts by breaking down a given problem

© Springer-Verlag Berlin Heidelberg 2016
N.T. Nguyen et al. (Eds.): TCCI XXIV, LNCS 9770, pp. 134–152, 2016.
DOI: 10.1007/978-3-662-53525-7_8

into sub-problems of the same type (*divide* phase), solving them separately (*conquer* phase) and merging the local results into a final solution. Although in the context of time series forecasting the term Divide and Conquer is not very used, the idea of splitting time series into subseries to increase the prediction accuracy, is fairly common. This approach is applied in time series clustering, fuzzy clustering, segmentation, information granules, knowledge discovery, discretization and even Fast Fourier transformation. In the seminal book "Data Mining: Concepts and Techniques" by Morgan Kaufmann. (Han 2001) five major categories of clustering are discussed: partitioning methods (e.g. k-mean algorithm MacQueen 1967), hierarchical methods (e.g. Chameleon algorithm Karypis 1999), density based methods (e.g. DBSCAN algorithm Ester 1996), grid-based methods (e.g. STING algorithm Wang 1997) and model-based methods (e.g. AutoClass algorithm Cheeseman 1996). Other data segmentation procedures includes time-series segmentation (Keogh 2004, Kovai 1995), granulation (1979, Pedrycz 2001, Wang 2015), knowledge discovery (de Boor 1978, Hppner 2002, Ramsay and Silverman 1997, JF 1983) and SAX algorithms Jessica Lin 2007.

The majority of these methods rely upon grouping elements within intervals, although this approach doesn't allow to identify patters spread across the whole timeseries. The proposed approach overcome this issue by decomposing the original problem into simpler sub-problems and applying the *divide* function evaluated during the training phase. The existence of a deterministic pattern within the original time series is initially assumed. The method is trained on the generated subseries implying the potential reduction of the prediction error with respect to the original time series (*divide* phase). Elements which are not included in the chosen subseries are grouped into a complementary subseries. Due to the deterministic nature of the *divide* function we know to which subseries the predicted elements belong to.

Currently there are many popular and well developed methods for time series forecasting such as ARIMA models, Neural Networks and Fuzzy Cognitive Maps (Makridakis 1997, Han 2003, Song and Miao 2010). Our ensemble approach leverages the prediction capabilities od such methods. Initially these forecasting methods can be used to evaluate the *divide* function (*divide* phase) and predict the values of every subseries separately (*conquer* phase). The next phase consists in merging the previously predicted values in order to obtain the prediction on the whole time-series (*merge* phase).

This paper is organized as follows. The proposed approach is described in detail in Sect. 2. The computational complexity of the algorithm overhead (according to the time series' length) is estimated in Sect. 3. The simulations performed on different series are presented in Sect. 4. The last Sect. 5 concludes the paper.

2   Methods

Any discrete time series can be decomposed into a finite number of subseries, each including a finite set of data points. The majority of the known methodologies

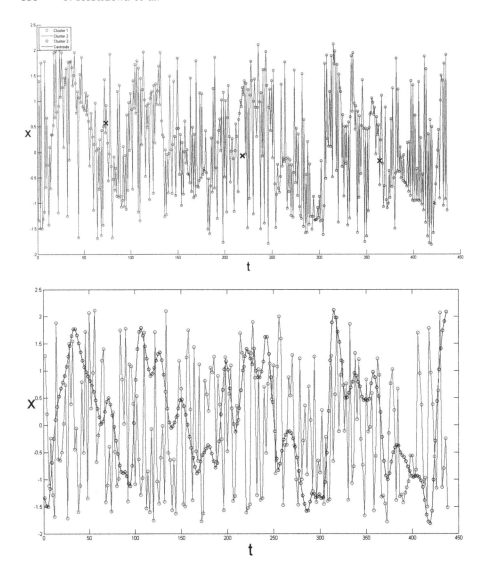

Fig. 1. The time series above was created by perturbing with random noise each second element of the IceTarget time series (Fig. 6). The k-mean clustering (top figure) generates subseries by splitting the original time series into intervals. The proposed approach (bottom figure) generates subseries by applying the *divide* function which selects meaningful points from the entire series. The second approach is capable of identifying the original IceTarget patterns whereas the first method is not.

rely upon grouping elements within time intervals. By definition this strategy can uniquely find dependencies between points which are contained in the same interval, failing to recognize general patterns possibly embedded throughout the whole time series. Our approach tackle this problem by applying a *divide* function to original time series. Such function generates subseries by selecting elements scattered across the whole time series. Such crucial difference between our approach and the already existent state of the art methods, is appreciable in Fig. 1. In the presented method the selection of such points is performed by a deterministic function (Fig. 1) for a graphical representation. Our assumption is that there exist subseries which are easier to predict than the original data series. Hence we introduce a Divide and Conquer (D&C) method which combines the subseries predictions in order to optimize the whole time series forecasting. The schematic description of the methodology is briefly resumed in Fig. 2.

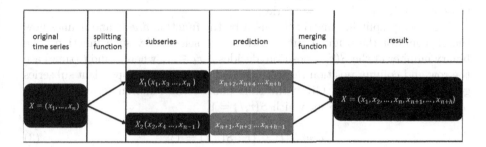

Fig. 2. Divide And Conquer method description

Our approach relies on finding a function named having the following properties:

1. It splits the vector X into subseries X_1 and its complement X_2 such that prediction error of that subseries would be lower then prediction error of time series X.
2. For any natural number t it assigns element at time t to the subseries X_1 or X_2.

After splitting X into subseries we use one of the previously mentioned methods (see Sect. 1) on the X_1 and X_2 subseries separately. Then we merge the results according to the above mentioned *divide* function. This procedure allows to perform a prediction on the original time series. The error is evaluated using the mean square error (MSE) metric.

$$MSE = \frac{1}{n} \sum_{i=1}^{n} (y - \hat{y})^2 \qquad (1)$$

where n is number of all predicted values, y is real value and $\hat{y}$ is the predicted value. In order to provide a clear description of our approach the algorithm is divided into simple functions which are described separately.

2.1 Divide Function Definition

Let $X = (x_1, x_2, ..., x_n)$ be a time series, then we can define:

$$\tau(x_i) = \{t_i \mid t_i \in \mathbb{N}\} \tag{2}$$

where x_i is the ith data point of the time series and t_i is the position in time associated with that data point. Given the time series $X = (x_1, x_2, ... x_n)$ we can for instance generate a new subseries x_{even} including only even elements. $x_{even} = (x_1^{even}, x_2^{even}, ..., x_{n/2}^{even})$. It is worth noticing that $x_i^{even} = x_{2i}$ so $\tau(x_i^{even}) = \tau(x_{2i})$ so $\tau(x_1^{even})$, $\tau(x_2^{even})$, ... , $\tau(x_{n/2}^{even}) = 2, 4, ..., n$. This is true because the τ function returns the time values associated with the original time series and not those associated to the new subseries.

The binary splitting matrix S is used by the function *divide* and defines how the elements of the original time series X are selected by such function. The binary number at the $S(r, t)$ position, decides if $x_t \in X_r$ where r and t (time) are the row and column position respectively and X_r is the correspondent subseries generated by the *divide*. Given the time point t the function returns the row positions of the matrix S for which $S(r, t) = 1$.

$$divide(t) = \{r \mid S(r, t) = 1\} \tag{3}$$

where S is proposed during the training process, r is the row number and t is time. Let the time series forecasting method be defined as:

$$x_{n+i} = time\_series\_forecasting(X, i) \tag{4}$$

where X is the time series $(x_1, x_2, ..., x_n)$ and x_{n+i} is i steps ahead the predicted value. For simplicity, let's consider the Divide and Conquer method with $i = 1$. Then our goal is to predict the first consequent value of the time-series.

$$x = time\_series\_forecasting(X_k, 1) \tag{5}$$

using

$$divide(n + 1) = m \tag{6}$$

where n is the time of the last element of the series and $n + 1$ is the time of the first predicted value, x is the predicted value of x_{t+1} and X_m is the mth subseries.

2.2 Divide Phase

2.2.1 Divide Series into Subseries with Matrix S.

Consider the time series $X = (x_1, x_2, ..., x_n)$ and let $S_m = (b_{m,1}, b_{m,2}, ..., b_{m,n})$ be the mth row of the binary matrix S. Let $X_1 = (x_1, x_2, ..., x_{n_1})$ be the subseries of all the elements x_i such that the corresponding $b_{1,i} = 1$. Analogously let X_2 be the subseries of all the elements x_i such that in the 2nd row of S, $b_{2,i} = 1$. For example:

$$X = (x_1, x_2, ..., x_n)$$

$$S = \frac{1\ 1\ 0\ 0\ 1\ 1\ ...\ 0\ 0}{0\ 0\ 1\ 1\ 0\ 0\ ...\ 1\ 1}$$

$$X_1 = (x_1, x_2, x_5, x_6, .., x_n)$$
$$X_2 = (x_3, x_4...x_{n-2}, x_{n-1})$$

2.2.2 Extend Binary Vector

Let s be a binary vector (row of the S matrix) to be extended for prediction purposes. The first step needed to extend the binary vector s consists in the generation of a dictionary including binary patterns. The pseudo-code of the function dedicated to this task is given in Table 2.

An empty dictionary is initialized. The interval of length $d + 1$ starting from the first position is generated. If the dictionary contains a binary pattern which coincides with a given pattern at the first d position but differs at the last position $(d + 1)$, then such pattern is too short to unequivocally extend the vector s. In such case the dictionary is cleared, d is incremented by 1 and the whole process is repeated starting from the first position. In case the dictionary does not contains the pattern we add it to the dictionary. Then the interval starting position is incremented by 1 and the whole process is repeat until the end of the interval don't exceed the end of the s vector. The function returns the dictionary when the end of the interval exceeds the end of the s vector (Fig. 3).

Once the dictionary has been computed the s vector can be further extended applying the *extend_binary_vector* function given in Table 3.

Such function takes both the binary vector s to be extended and the expected length of the new vector as input values. The *create_dictionary* function provides the dictionary for the binary vector s. Let d be length of every vector in the dictionary and n be the length of the vector s. In the dictionary we search for a pattern having $S(b_{n-d+2}...b_n)$ excluding the last position. If there is no such pattern we extend s by including a random binary number. Otherwise we extend s by the last value of the pattern found within the dictionary. We repeat this process until s reaches the expected *new_length* (Table 1).

Table 1. Pseudo-code of the subseries generation according to the vector s

```
FUNCTION create_subseries(s,X)
subseries_X = []
FOR i=1;i++;i<=size(s)
   IF s(i)==1
      subseries_X = subseries_X.add(X(i))
   ENDIF
ENDFOR
RETURN subseries_X
```

Table 2. Pseudo-code of the dictionary generation for a given pattern *s*

```
FUNCTION create_dictionary(s)
d=1
dict = []
i=1
WHILE i<=(size(s)-d)
   window = s(i:i+d)
   IF ismember(window(1:end-1),dict(1:end-1)
      !ismember(window,dict) )
      d=d+1
      i=1
      dict = []
   ELSE
      dict = dict.add_new_row(window)
   ENDIF
   i = i+1
ENDWHILE
RETURN dict
```

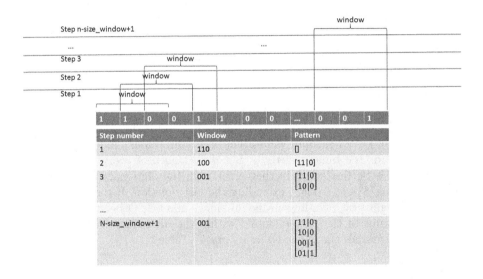

Fig. 3. Create dictionary for binary vector

2.2.3   Find Proposition of Best Subseries

A pseudo-code definition of the algorithm for finding the best subseries proposal is given in Table 4.

The *find_best_subseries* function accepts a time series, an arbitrary chosen constants c and the multiplicity number k. All the possible binary vectors of length c are generated under the condition that the total number of 1s within c is equal or greater then $\lceil \frac{c}{2} \rceil$.

Table 3. Pseudo-code of algorithm which extends the binary vector

```
FUNCTION extend_binary_vector(s,new_length)
dict = create_dictionary(s)
i = size(s)-length_of_row(dict)+2
WHILE length(s)<new_length
    small_window = s(i:i+length_of_row(dict)-2)
    index = index_of_element(small_window,dict(:,1:end-1) )
    IF index>0
        s.add(dict.elementAt(index).elementAt(end))
    ELSE
        s.add(randomly_0_or_1())
    ENDIF
    i = i+1
ENDWHILE
RETURN s
```

These vectors are initialized as rows of the matrix S. Hence S has m rows where $m = 2^{c-1}$ for odd c values and $m = 2^{c-1} + \frac{1}{2}\binom{c}{c/2}$ for even c values. For every i^{th} row of the S matrix a vector X_i is generated, as described in Sect. 2.2. Each subseries X_i^1 is split into the training set $X_{i_{train}}$ and the test set $X_{i_{test}}$ including 70 % and 30 % of the series, respectively. The values of the train-to-test ratio were chosen arbitrarily. After the application of an arbitrary chosen prediction method (e.g. NN, ARIMA, Fuzzy Cognitive Maps etc.), the MSE is computed in order to evaluate the prediction performances of the method. It worth noticing that the process can be computed in parallel on the vectors $X_1, X_2, ..., X_m \in X$. This approach can significantly reduce the computational time required for the prediction. The number of X_i subseries increase exponentially with respect to c making this phase the most computationally expensive part of the algorithm presented. This part could be further optimized in order to improve the performances of the method.

Let consider the set of pairs (S_1, MSE_1) , (S_2, MSE_2) , (S_3, MSE_3) ... (S_m, MSE_m) where S_i is i^{th} row of the matrix S. We can reject all the rows of S but $\lceil \frac{1}{k} \rceil$ of the rows with the lowest MSE. We extend the rows of S using the function *extend_binary_vector* (refer to pseudo-code in Table 3) to obtain the binary vectors with length $k \cdot c$. We obtain the matrix S of dimension $[\lceil \frac{m}{k} \rceil \times k \cdot c]$ so that every row $S_1, S_2, ..., S_{\lceil \frac{m}{k} \rceil}$ has length $k \cdot c$. For every row S_i we create the vectors X_1 as previously described in Sect. 2.2. We repeat the MSE calculations, selecting and extending S until its rows exceed the length of the training set. The row S_i with the corresponding lowest MSE value (Fig. 4) is then returned by the function as a result. At the next step we create a matrix such that the first row is a vector returned by the *find_best_subseries* function and second row is the complementary vector with respect to the first row. Hence X can be divided into X_1 and X_2.

Table 4. Pseudo-code of algorithm which finds proposition of best subseries

```
FUNCTION find_best_subseries(time_series,c,k)
S = cob(c)//cob returns all binary combinations
//of length c with 1 on at least c/2 positions
FOR i=1;i++;i<=number_of_rows(S)
    X1(i,:)=create_subseries(S(i,:),time_series)
    Xtrain=X1(1:0.7*size(X))
    Xtest =X1(0.7*size(X):end)
    MSE = chosen_prediction_method(Xtrain,Xtest)
    S(i,end+1) = MSE
ENDFOR
S = sort_ascendning_by_last_column(S)
S = S(1:ceiling(end/k),:)
WHILE c<size(time_series)
    c=k*c
    IF c>size(time_series)
        c = size(time_series
    ENDIF
    FOR j=1;j++;j<number_of_rows(S)
        S(j,:)=extend_binary_vector(S(j,:),c)
        X(j,:)=create_subseries(S(j,:),time_series)
        Xtrain=X(1:0.7*size(X))
        Xtest=X(0.7*size(X):end)
        MSE=any_prediction_method(Xtrain,Xtest)
        S(j,end+1) = MSE
    ENDFOR
    S = sort_ascendning_by_last_column(S)
    S = S(1:ceiling(end/k),:)
ENDWHILE
//return s with lowest MSE
RETURN S(1,:)
```

2.3   Conquer Phase

During the *conquer* phase the subseries are predicted separately. The same predictive method used in the *divide* phase can be applied, however any other forecasting method can be exploited. For this reason the D&C method approach can be defined as an *ensemble* method. It is important to consider that although the process of the S matrix evaluation is time consuming (exponential time complexity with respect to the c parameter as mentioned in Sect. 3), the subseries prediction has almost no overhead. Hence, while it is recommended to apply a fast and deterministic methods to the *divide* function, much more time consuming and accurate prediction methods can be used during the *conquer* phase.

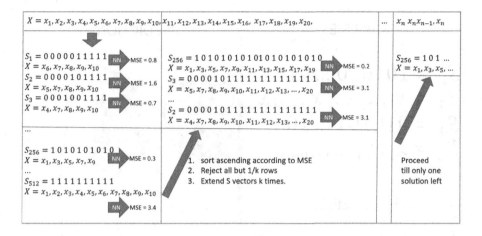

Fig. 4. Schematic view of the algorithm proposing the best subseries

2.4 Merge Phase

In order to predict the value of x_{t+1} we need to predict the $(t+1)^{th}$ column within
the S matrix (Table 3). If $S(1,t+1) = 1$ than we select the prediction calculated
on the subseries X_1 otherwise we choose the prediction from the subseries X_2,
where X_2 is the complementary subseries X_1 with respect to X.

3   Complexity of Proposed Approach

Our goal is to show that our approach has time complexity equal to
$\mathcal{O}(log_k^n MSE(n))$, where MSE is an arbitrary chosen prediction function, k is
the constant multiplicity parameter and n is the series length. We assume that
the prediction function has a complexity greater then $\mathcal{O}(n)$. The complexity of
each function of the algorithm is computed separately and presented below.

3.1   Time Complexity of the *create_subseries* Function Corresponding Subseries

The algorithm creates vectors X_1 and X_2 from the original series using m the s
is described in the Sect. 2.2. The complexity of the algorithm is $\mathcal{O}(n)$ (Fig. 5).

3.2   Time Complexity of the *extend_binary_vector* Function

This algorithm extends the binary vector (refer to pseudo-code in Table 3) and is
divided into two parts. Firstly it creates the dictionary needed for the extension
of the binary sequence. We can notice that the maximal number of vectors in
the dictionary is greater then 2^d where d is the length of the vector. However,
at the same time the dictionary cannot contain more then c elements where c

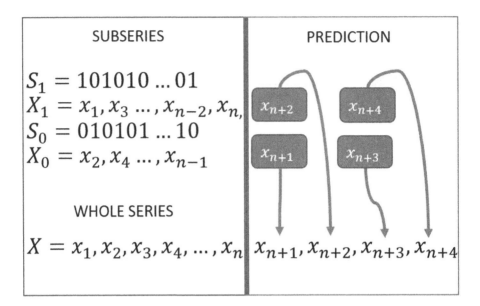

Fig. 5. Merge process description

is the length of the vector on which we construct the dictionary. Therefore, for each given step, the dictionary length is smaller than $min(2^d, c)$. Moreover we know that the algorithm will produce no more that c such dictionaries. Since the number of operation is equal to

$$\sum_{i=1}^{c} min(2^d, c) * c < c^2 \qquad (7)$$

we can imply that the time complexity of such algorithm is $\mathcal{O}(c^2)$. Another part of the algorithm foresee the extension of the binary vector using the dictionary previously prepared. The dictionary is computed only once and it is then used during the whole prediction process. Finding a specific vector in a dictionary doesn't cost more than $\mathcal{O}(log_2 c)$. For these reasons extending the binary vector by n elements costs $\mathcal{O}(nlog_2 c)$.

3.3   Time Complexity of the algorithm *finds_best_subseries*

We choose some arbitrary length of the first subseries c and the multiplicity parameter k. We start with subseries of length c and then in every step we extend this subseries k times. We also removes all the rows of S but $\lceil \frac{1}{k} \rceil$. The number of operations can be approximated by:

$$2^{c-1}MSE(c) + 2^{c-1}c^2 + clog_2(c) +$$
$$\lceil\frac{1}{k}\rceil 2^{c-1}MSE(kc) + kclog_2(c) + \lceil(\frac{1}{k})^2\rceil 2^{c-1}MSE(k^2c) +$$
$$k^2clog_2(c) + ... + \lceil(\frac{1}{k})^{log_k^{n/c}}\rceil 2^{c-1}MSE(n)$$

which is equal to

$$2^{c-1}c^2 + \sum_{i=0}^{log_k^{n/c}} \lceil(\frac{1}{k})^i\rceil 2^{c-1}(MSE(k^ic) + k^iclog_2c) \tag{8}$$

where 2^{c-1} is number of all proposals of S proposed in the first step, c^2 is maximal cost for creation binary vector dictionary (refer to Sect. 3.2), $log_k^{n/c}$ is the maximal number of steps after which length of S reaches n, $MSE(k^ic)$ is cost of approximation prediction error on every step for every S proposal, k^iclog_2c is cost of extending binary vector S k times. Taking into account this equation we can say that the number of operation considered in our approach is definitely smaller than

$$2^cc^2 + log_k^{n/c}(2^c)(MSE(n) + (nlog_2c)) \tag{9}$$

Taking into consideration that the complexity of $MSE(n)$ is not less than $\mathcal{O}(n)$ and omitting constant operations, we can affirm that the complexity of our approach with respect to n can be approximated to:

$$\mathcal{O}(log_k^n MSE(n)) \tag{10}$$

On the contrary, the time complexity with respect to c can be approximated to:

$$\mathcal{O}(2^c) \tag{11}$$

It is worth noticing that algorithm can be processed in parallel and consequently the time of computation can decrease significantly.

4   Simulations

In order to check the efficiency of our approach we made several simulations. In our simulations we used neural networks (NN) with one hidden layer and a delay value equal to 2 (refer to diagram on Fig. 7). As a neural network training method we used Levenberg-Marquardt back-propagation algorithm (Marquardt 1963). During the simulations comparative to our method, we used neural networks with the same structure, training rate, training method and number of iterations with respect to other methods. The only difference was that while the Neural Networks used in our approach were trained on subseries chosen by our algorithm, the NN used in the comparative simulations were trained on the whole training set. In every simulation the constant number $c = 12$ and the multiplicity parameter $k = 2$ were used. In order to avoid random bias we repeated every simulation

10 times and reported the mean results. We also used IceTargets data which contains a time series of 219 scalar values representing measurements of global ice volume over the last 440,000 years (Fig. 6). Time series is available at [http://lib.stat.cmu.edu/datasets/], or in the standard Matlab library as *ice_dataset*.

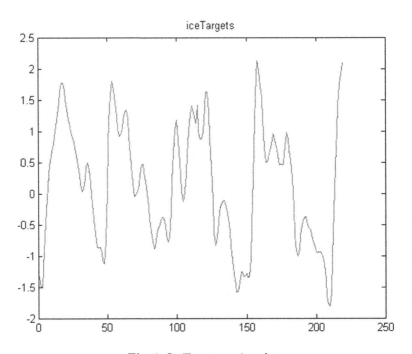

Fig. 6. IceTargets series plot

4.1   IceTargets with Random Noise

We modified the *IceTargets* series by adding random numbers generated from a uniform distribution upon the interval $[-1.81, 2.12]$, where the extremes of the interval represents the minimum and maximum values of the *IceTargets* series, respectively. The pseudo-random number occurrence scheme is defined by the following relation:

X = (rand(1),*IceTargets*(1),rand(2),*IceTargets*(2), ... *IceTargets*(219),rand (220))

Our approach produces the matrix:

$$S = \begin{matrix} 0\ 1\ 0\ 1\ 0\ 1\ 0\ 1 \\ 1\ 0\ 1\ 0\ 1\ 0\ 1\ 0 \end{matrix}$$

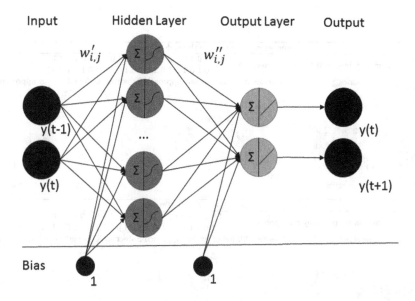

Fig. 7. Diagram of neural network used in simulations

Table 5. Comparisons with other methods for time series based on MSE.

	IceTargets merged with noise	*IceTargets* merged with cos	*Quarterly Gross Farm Product*
Our approach	0,62	0,0170	0,007
Single Neural Network	0,93	0,5144	0,0211
Increase efficiency	1,5 times	30,25 times	3,014 times

which was found to identify the correct pattern, splitting the time series according to it. For this reason, NN are able to predict *IceTargets* series and random noise separately. Our approach displayed a mean MSE value equal to 0.62 while the neural networks trained on the whole dataset shows a MSE value equal to 0.93. The results are presented in Table 5.

4.2   Cosinus with IceTargets

We created time series by merging cosinus and *IceTargets* time series using pattern:

$X = (\cos(0.1),\ IceTargets(1),\ \cos(0.2),\ \cos(0.3),\ IceTargets(2),\ IceTargets(3),\ \cos(0.4),\ IceTargets(4),\ \cos(0.5),\ \cos(0.6),\ IceTargets(5),\ IceTargets(6)\ ...)$

The pattern can be described by matrix:

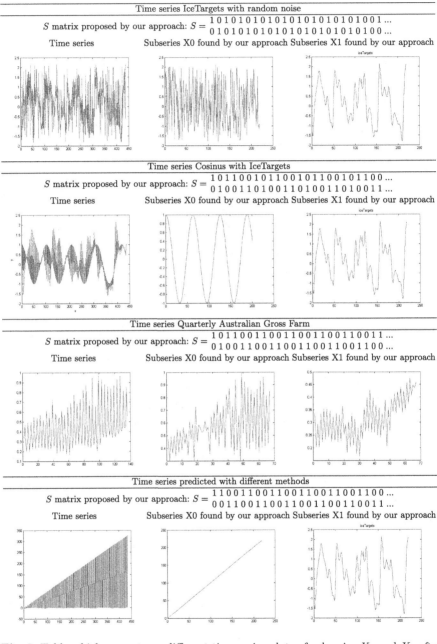

Fig. 8. Table which presents on different time series plots of subseries X_1 and X_2 after divide with our approach

$$S = \frac{1\ 0\ 1\ 1\ 0\ 0\ 1\ 0\ 1\ 1\ 0\ 0\ 1\ 0\ 1\ 1\ 0\ 0\ 1\ 0\ 1\ 1\ 0\ 0 \dots}{0\ 1\ 0\ 0\ 1\ 1\ 0\ 1\ 0\ 0\ 1\ 1\ 0\ 1\ 0\ 0\ 1\ 1\ 0\ 1\ 0\ 0\ 1\ 1 \dots}$$

Our approach is able to identify the correct pattern by splitting the time series into the expected subseries (Fig. 8). Neural networks trained on the subseries shows $MSE = 0,0170$, while neural network trained on whole training set gives $MSE = 0,5144$.

4.3 Quarterly Australian Gross Farm Product

In this simulation we used the real statistic data of the Quarterly Australian Gross Farm Product $m 1989/90 prices. The time series is constituted by 135 data points representing the values measured between September 1959 and March 1993. These data are available at (Australian Bureau of Statistics Australian Bureau of Statistics [2015]). The data were normalized to a 0-1 interval. The average MSE value calculated using our approach was equal to 0,007 while the average value of MSE achieved by a single neural network was equal to 0,0211 (refer to Table 5).

4.4 Series Predicted with Different Methods

In all previous simulations we used our approach to transform time series into subseries and then we predict their values with the same method - neural network. However, our approach gives the possibility to use different methods of prediction on each subseries. For this reason we can choose different methods according to specific prediction properties of each subseries and take advantage of such methods. In order to display such possibility we merged two series with completely different prediction properties into one time series. We choose a simple series which grows linearly with respect to time but doesn't seems to change in time with respect to the $IceTargets$ statistic data. Since such series show a mixed behaviour influenced by the different nature of its constitutive components, we can laverage the predicting capabilities of different methods which act optimally on each given component. The subseries were merged conforming to the following pattern:

$X = (1, 2, IceTargets(1), IceTargets(2), 3, 4, IceTargets(3), IceTargets(4),$
$5, 6, IceTargets(5), IceTargets(6) \dots)$

Such pattern is described by the matrix

$$S = \frac{1\ 1\ 0\ 0\ 1\ 1\ 0\ 0\ 1\ 1\ 0\ 0\ 1\ 1\ 0\ 0\ 1\ 1\ 0\ 0\ 1\ 1\ 0\ 0 \dots}{0\ 0\ 1\ 1\ 0\ 0\ 1\ 1\ 0\ 0\ 1\ 1\ 0\ 0\ 1\ 1\ 0\ 0\ 1\ 1\ 0\ 0\ 1\ 1 \dots}$$

The algorithm divides the original time series into two subseries (Fig. 8). Exploiting the different properties of such subseries we selected linear regression and neural network to predict X_1 and X_2, respectively. This strategy allowed to lower the MSE value down to 0.0101. In comparison, the application of a single neural network method displayed an average $MSE = 2535.45$ while the usage of a single linear regression method showed $MSE = 30.35$ (Table 6) This example exhibit the potential benefit of the ensemble approach proposed with respect to single predictors.

Table 6. Comparison of MSE calculated with different methods for the time series created by merging linear function and IceTarget.

Method	Neural Network	Linear regression	Our approach
MSE	2535.45	30.35	0,0101

5   Conclusions

In presented work, we proposed a novel method for time series forecasting. Our approach is based on dividing the series into a subseries and its complement, predicting their values separately and then merging the prediction results in the final prediction. This strategy can lower the potential prediction error with respect to the prediction based on whole set. Moreover it allows the application of different prediction methods to both subseries, combining their respective benefits. The proposed approach is not associated with any specific time series forecasting method and can be applied as a generic solution for several time series pre-processing problems. We show that our approach is capable of performing noise filtering. In order to validate the efficiency of the introduced solution we conducted a series of experiments. The results obtained showed a significant improvement of the prediction accuracy of the ensemble method with respect to the respective base learners. Moreover we have shown that the overhead generated by the algorithm is asymptotically logarithmic with respect to the length of the time series. The computations can be processed in parallel decreasing the computational time required for the forecasting.

Our solution opens up broad prospects of further work. In the current implementation the method divide series into exactly two subseries. It would be worth investigating the division into multiple subseries. Furthermore the elements belonging to such generated subseries could be sampled simultaneously and combined into new prediction proposals. The impact of the c parameter and the minimal acceptable subseries length also need further investigation. Wide scale studies on real data could help to identify the optimal field of application for the proposed method. One of future area of research could also include the design and implementation of automated procedures targeted to the selection of different prediction methods to be applied on the proposed subseries.

We believe that the algorithm here presented could give a contribution to the field of time series forecasting.

Acknowledgements. This research was supported by the European Union from financial resources of the European Social Fund, Project PO KL Information technologies: Research and their interdisciplinary applications and by the Polish National Science Centre with the grants 2014/15/B/ST6/05082 and 2013/09/B/NZ2/00121.

References

Australian Bureau of Statistics. https://datamarket.com/data/set/22xn/quarterly-australian-gross-farm-product-m-198990-prices-sep-59-mar-93/, Accessed 19-July-2015

de Boor, C.: A practical guide to splines (1978)

Karypis, G., Han, E.H., Kumar, V.: Chameleon: hierarchical clustering using dynamic modeling. Computer **32**, 68–75 (1999)

http://lib.stat.cmu.edu/datasets/

Wu, H., Sharp, G., Salzberg, B., Kaeli, D., Shirato, H., Jiang, S.: Subsequence matching on structured time series data. In: SIGMOD (2005)

Hppner, F.: Knowledge discovery from sequential data (2002)

Han, J., Kamber, M.: Data mining: Concepts and techniques. Morgan Kaufmann, San Francisco (2001)

Han, J., Kamber, M.: Application of neural networks to an emerging financial market: forecasting and trading the taiwan stock index. Comput. Oper. Res. **30**, 901–923 (2003)

Lin, J., Keogh, E., Wei, L., Lonardi, S.: Experiencing sax: a novel symbolic representation of time series. Data Min. Knowl. Discov. **15**(2), 107–144 (2007)

JF, A.: Maintaining knowledge about temporal intervals, pp. 832–843 (1983)

Keogh, E.: A survey and novel approach, pp. 1–22 (2004)

Kovai, Z.: Time series analysis, faculty of economics (1995)

La, Z.: Fuzzy sets and information granularity, pp. 3–18 (1979)

Ester, M., Kriegel, H.-P., Jiirg, S., Xiaowei, X.: A densitybased algorithm for discovering clusters in large spatial databases. In: Proceedings of the 1996 International Conference on Knowledge Discovery and Data Mining (KDD 1996) (1996)

MacQueen, J.: Some methods for classification and analysis of multivariate observations. In: Proceedings of 5th Berkeley Symposium on Mathematical Statistics and Probability 1, pp. 281–297. University of California Press (1967)

Marquardt, D.: An algorithm for least-squares estimation of nonlinear parameters. SIAM J. Appl. Math. **11**(2), 431–441 (1963)

Cheeseman, P., Stutz, J.: Sting: a statistical information grid approach to spatial data mining. Bayesian classification (AutoClass): theory and results. In: Fayyard, U.M., Piatetsky-Shapiro, G., Smyth, P., Uthurusamy, R. (eds.) Advances in Knowledge Discovery and Data Mining. AAAI/MIT Press, Cambridge, MA (1996)

Pedrycz, W., Vukovich, G.: Abstraction and specialization of information granules, pp. 106–111 (2001)

Ramsay, J.O., Silverman, B.W.: Functional data analysis (1997)

Makridakis, S., Wheelwright, S., Hyndman, R.: Forecasting: Methods and applications. Wiley, New York (1997)

Song, H.J., Shen, Z.Q., Miao, C., Miao, Y.C.: Fuzzy cognitive map learning based on multi-objective particle swarm optimization. IEEE Trans. Fuzzy **18**(2), 233–250 (2010)

Tong, H.: Threshold models in non-linear time series analysis. Springer, Heidelberg (1983)

Wang, W., Yang, J., Reeves, M.R.: Sting: a statistical information grid approach to spatial data mining. In: Proceedings of the 1997 International Conference on Very Large Data Base (VLDB 1997) (1997)

Wang, W., WitoldPedry, X.L.: Time series long-term forecasting model based on information granules and fuzzy clustering, pp. 17–24 (2015)

Zhang, G.: Time series forecasting using a hybrid arima and neural network model. Neurocomputing **50**, 159–175 (2003)

Zhang, G.: A neural network ensemble method with jittered training data for time series forecasting. Inf. Sci. **177**, 5329–5346 (2007)

Application Areas of Ephemeral Computing: A Survey

Carlos Cotta[1], Antonio J. Fernández-Leiva[1], Francisco Fernández de Vega[2], Francisco Chávez[3(✉)], Juan J. Merelo[4], Pedro A. Castillo[4], David Camacho[5], and María D. R-Moreno[5]

[1] Dept. Lenguajes Y Ciencias de la Computación,
Universidad de Málaga, Malaga, Spain
{ccottap,afdez}@lcc.uma.es
[2] Dept. Tecnología de los Computadores y de las Comunicaciones,
Universidad de Extremadura, Merida, Spain
fcofdez@unex.es
[3] Dept. Ingeniería En Sistemas Informáticos Y Telemáticos,
Universidad de Extremadura, Merida, Spain
fchavez@unex.es
[4] Dept. Arquitectura Y Tecnología de los Computadores,
Universidad de Granada, Granada, Spain
{jmerelo,pacv}@ugr.es
[5] Dept. Ingeniería Informática, Universidad Autónoma de Madrid, Madrid, Spain
{david.camacho,gema.bello}@uam.es

Abstract. It is increasingly common that computational devices with significant computing power are underexploited. Some of the reasons for that are due to frequent idle-time or to the low computational demand of the tasks they perform, either sporadically or in their regular duty. The exploitation of this (otherwise-wasted) computational power is a cost-effective solution for solving complex computational tasks. Individually (device-wise), this computational power can sometimes comprise a stable, long-lasting availability window but it will more frequently take the form of brief, ephemeral bursts. Then, in this context a highly dynamic and volatile computational landscape emerges from the collective contribution of such numerous devices. Algorithms consciously running on this kind of environment require specific properties in terms of flexibility, plasticity and robustness. Bioinspired algorithms are particularly well suited to this endeavor, thanks to some of the features they inherit from their biological sources of inspiration, namely decentralized functioning, intrinsic parallelism, resilience, and adaptiveness. Deploying bioinspired techniques on this scenario, and conducting analysis and modelling of the underlying Ephemeral Computing environment will also pave the way for the application of other non-bioinspired techniques on this computational domain. Computational creativity and content generation in video games are applications areas of the foremost economical interest and are well suited to Ephemeral Computing due to their intrinsic ephemeral nature and the widespread abundance of gaming applications in all kinds of devices. In this paper, we will explain why and how they can be adapted to this new environment.

© Springer-Verlag Berlin Heidelberg 2016
N.T. Nguyen et al. (Eds.): TCCI XXIV, LNCS 9770, pp. 153–167, 2016.
DOI: 10.1007/978-3-662-53525-7_9

Keywords: Ephemeral computing · Bioinspired optimization · Evolutionary computation · Complex systems · Autonomic computing · Distributed computing

1   Ephemeral Computing: What and Why

This paper revolves around the notion of *Ephemeral Computing* (Eph-C) which can be defined as *"the use and exploitation of computing resources whose availability is ephemeral (i.e., transitory and short-lived) in order to carry out complex computational tasks"*. The main goal in Eph-C is thus making an effective use of highly-volatile resources whose computational power (which can be collectively enormous) would be otherwise wasted or under-exploited. Think for example of the pervasive abundance of networked handheld devices, tablets and, lately, wearables –not to mention more classical devices such as desktop computers– whose computational capabilities are often not fully exploited. Hence, the concept of Eph-C partially overlaps with ubiquitous computing, volunteer computing and distributed computing. Due to these research fields deal with the concept of ephemerality are explained in next section, but exhibits its own distinctive features, mainly in terms of the extreme dynamism of the underlying resources, and the ephemerality-aware nature of the computation which autonomously adapt to the ever-changing computational landscape. These concepts not trying to fit to the inherent volatility of the latter but even trying to use it for its own advantage.

In light of the computational context described above, it is clear that the algorithmic processes deployed onto it should be flexible (to work on a variety of computational scenarios), resilient (to cope with sudden failures and with the phenomenon of churn [55]), (self-)adaptive (to react autonomously to changes in the environment and optimize its own performance in a smart way), and intrinsically decentralized (since centralized control strategies cannot consistently comprehend the state of the computational landscape and decisions emerging from them would lag behind the changing conditions of the latter). Some bioinspired algorithms, like the Evolutionary algorithms (EAs) fit nicely into this scenario. However, few works have previously considered the interest of endowing evolutionary algorithms with the capability for coping with transient behaviors in underlaying computer systems. Moreover, in the age when the term Big Data [35] is present in many initiatives requiring large amount of computational resources for storing, processing, and learning from huge amount of data, new methods and algorithms for properly managing heterogeneous computing resources widely distributed along the world are required. Energy consumption must also be considered from the point of view of both algorithms and hardware resources, given the large differences among large computing infrastructures typically devoted to running massively parallel algorithms when compared to smart devices optimized for reducing battery consumption. It is of the foremost interest to research on the basic features allowing to provide efficient and reliable ephemeral evolutionary services.

The paper is structured as follows. Section 2 gives an onverview of Eph-C. Then, we analyze an important parameter in Eph-C: the energy consumption.

Next, the optimization criterion is presented by means of bioinspired algorithms. Section 5 presents a short revision on bioinspired methods and applications that could be affected by Eph-C characteristics. Finally, the conclusions are outlined.

2   Ephemeral Computing in Perspective

According to the Oxford Dictionary, the term *ephemeral* means "lasting for a very short time". It thus encompasses things or events with a transitory nature, with a brief existence. A number of phenomena and resources in computer science are endowed with that feature (e.g. in computing networking, an ephemeral port is a TCP port, for instance, dynamically assigned to a client application for a brief period of time, in contrast with well known ports) [10]. Ephemeral behaviors can be also observed in the way users collaborate in volunteer networks of computers.

Although ephemeral phenomena naturally arise in several areas such as ubiquitous computing, volunteer computing or traditional research areas like distributed computing, some issues arise when dealing with ephemeral behavior. In cloud computing [3], for instance, the opposite is usually looked for: *persistence*. Although services are commonly associated with computations among autonomous heterogeneous parties in dynamic environments, exceptions must be handled to take corrective actions. Ephemeral services are thus commonly seen more as a problem than a solution [28].

On the other hand, in ubiquitous computing the main goal is to leverage computation everywhere and anywhere, so that computation can occur using any kind of device, in any location, starting and ending at any time and using any format and during any amount of time. The main efforts in this area have been oriented to design and develop the underlying technologies needed to support ubiquitous computing [37] (like advanced middleware, operating systems, mobile code, sensors, microprocessors, new I/O and user interfaces, networks or mobile protocols). However, and in the same way it happens with cloud computing, the main target in ubiquitous computing is to allow stable and persistent computation processes perform a complete execution of the programs. When this area handles the concept of ephemeral devices, services or computation, the main solution is to stop the process, or processes, and resume once new devices are ready [59]. Previous hypothesis and assumptions can be extrapolated to distributed computing, where the concept of ephemeral services can be a problem that could eventually generate a failure in the execution of the process [54].

As stated before, the main focus of Eph-C is different from the above approaches: rather than trying to build layers onto the network of ephemeral resources in order to "hide" their transient nature and provide the illusion of a virtual stable environment, Eph-C applications are fully aware of the nature of the computational landscape and are specifically built to live (and optimize their performance) in this realm. Note that this does not imply the latter have a lower-level vision of the underlying computational substrate, or at least not markedly so. In fact, most low-level features can be abstracted without precluding attaining a more accurate vision of this fluctuating substrate.

To some extent, some of these ephemerality issues are also present in areas such as volunteer computing (VC) [53], whereby a dynamic collection of computing devices collaborate in solving a massive computational task, decomposing it into small processing chunks. Most VC approaches follow a centralized master/slave scheme though, and typically deal with resource volatility via redundant computation. A much more decentralized, emergent approach can be found in amorphous computing [1], but that paradigm is more geared towards programmable materials and their use to attack massive simulation problems. Massive problems are also the theme in ultrascale computing, where issues such as scalability, resilience to failures, energy management, and handling of large volume of data are of paramount importance [30,46]. Note however that Eph-C is not necessarily exascale nor it is oriented towards supercomputing.

3   Energy Consumption of Algorithms

When dealing with Eph-C, a relevant parameter to be taken into account is energy consumption, given that mobile devices frequently provide hardware resources to run programs, and battery consumption is always a concern for this kind of devices. During the last decades, researchers have focused on energy consumption for some kind of algorithms; for instance, encryption algorithms, that are frequently required in wireless communication, consume significant amount of computing resources such as CPU time, memory, and battery power, which affects every mobile device with wireless connection [51]. Sorting algorithms have been also analysed from this point of view [11].

Researchers have been interested in finding a proper way to map energy consumption to program structure [18]. But not always this mapping can be easily found, particularly when dealing with stochastic algorithms. The focus has been typically placed on the infrastructures behind the algorithms and the way they are exploited and offered to companies, which is the case for cloud models [8]; and not so frequently on the way algorithms can be optimized to better exploit those infrastructures. Thus, the term *Green Computing* has emerged when refering to the practice of using computing resources more efficiently while maintaining or increasing overall performance [26].

But we are more interested in the relationship between algorithms and the time and energy required to solve a problem. Although initially we could find a direct relationship between CPU cycles and energy required to run an algorithm, when optimization problems are faced by means stochastic algorithms affected by a series of parameters, such a relationship is not so straightforward, especially when different hardware architectures and operating systems can be chosen to run the algorithms: x86 family based computers running Linux or Windows, ARM based mobile devices with Android, etc. Moreover, even in a single platform and a given algorithm, different data structures that could be employed within the algorithm also affect energy consumed by some physical components of the computer, such as cache memories, and therefore can also be optimized to save energy, even when the time required to run an algorithm may not change [2].

If we specifically deal with ubiquitous computing, and given that the computation can occur using any kind of device, in any location, energy consumption should be considered when deciding which of the available devices will be employed. But also in a more standard setting, total energy consumed on a given hardware platform could in the future decide which is the preferred one, regardless of the time required to run the algorithm: sometimes the investment applied when finding a solution could be more importantly considered than time to solution.

4 Bioinspired Algorithms and Ephemerality

The term *bioinspired algorithms* usually refers to methods that draw some inspiration from Nature to solve search, optimization or pattern recognition problems. If we focus on optimization problems, the most prominent bioinspired paradigms are evolutionary computation and swarm intelligence. We are particularly interested in this kind of population-based search and optimization algorithms, which have a natural path to distributed computing by simply distributing the population among the different computing nodes, the issue being how to do it in an algorithmically efficient and scalable way. Eph-C, besides the obvious fact that the contribution of a node might come and go at any time, adds several other dimensions to the design of algorithms:

- *Inclusion*: all nodes should have a meaningful contribution to the final result, and they should be incorporated to the distributed system in such a way that they do.
- *Energy Consumption*: Different algorithm parameters influences time to result which in combination with specific hardware architectures behind each node implies a given energy consumption.
- *Asynchrony*: nodes communicate with the others without a fixed schedule due to their different performance.
- *Resilience*: the sudden disappearance of computing nodes must not destabilize the functioning of the algorithm.
- *Emergence*: the nature of the computational environment does not allow a centralized control and requires decentralized, emergent behavior.
- *Self-adaptation*: the algorithm should adapt itself to the changing computational landscape.

This latter issue is particularly important, and encompasses a number of self-⋆ properties [4] the system must exhibit in order to exert advanced control on its own functioning and/or structure, e.g., self-maintaining in proper state, self-healing externally infringed damage [20], self-adapting to different environmental conditions [48], and even self-generating new functionalities just to cite a few examples, see also [14,16]. Quite interestingly, these properties are frequently intrinsic features of the system, that is, emergent properties of its complex structure, rather than the result of endowing it with a central command. This also implies there is no need for a central control in the system; every node schedules itself. This decentralization implies a certain fault-tolerance due to the lack of a single point of failure, but it also means resilience must be built into the algorithms present in each node so that their sensitivity to changes in the rest of the system is minimal. This will include measures such as population sizing and the conservation of diversity in each node, as indicated by Cantú-Paz in [13] but taken to new meanings in this context. Indeed, models and algorithms have to be designed to be fault-tolerant [47] so that inclusion of new nodes will be done in a self-adaptive way, but also in such a way that its disappearance from the network will not have a big impact on performance. In fact, VC systems, which are an early example of Eph-C, have been proved to be fault tolerant to a certain point [22], but this fault tolerance will have to be taken into account not just at the implementation level (backing up solutions, for instance) but also at the model and algorithm level, measuring the impact of different churn models [33, 49].

Regarding Energy Consumption and Bioinspired Algorithms, we know that they have already been applied to optimize problems related to energy management and consumption, such as HAVC (heating, ventilating and air-conditioning) systems [19]. Yet, to the best of our knowledge, an analysis between the different flavors of available EAs, the parameters affecting them, and the relationship with different available configurations, time to solution, and energy consumed to reach the solution when different hardware architectures are employed have not been addressed. We think that this issue provides a new perspective to apply a multiobjective analysis of the algorithms considering time to solution, energy required, hardware architectures available in relatinship with algorithms configurations and main parameters. And this issue is particularly important when Eph-C is available.

5   Ephemeral Computing-Based Applications

From a practical perspective, the application, development and even deployment of any application that should be executed in an ephemeral environment will need to take into account those features described in Sect. 4. Therefore, the application to ephemeral environments of traditional techniques and methods, as bio-inspired computation, will generate some interesting challenges and opportunities that can be analyzed.

This section provides a short revision on some of those bioinspired methods and applications that could be affected by Eph-C characteristics.

5.1   Big Data and Bio-Inspired Clustering

The data volume and the multitude of sources have experienced an exponential growing with new technological and application challenges. The data generation has been estimated as 2.5 quintillion bytes of data per day[1]. This data comes

---

[1] http://www-01.ibm.com/software/data/bigdata/what-is-big-data.html.

from everywhere: sensors used to gather climate, traffic, air flight information, posts to social media sites (i.e. Twitter or Facebook as popular examples), digital pictures and videos (YouTube users upload 72 h of new video content per minute[2]), purchase transaction records, or cell phone GPS signals to name a few. The classic methods, algorithms, frameworks or tools for data management have become both inadequate for processing these amount of data and unable to offer effective solutions to deal with the data growing. The management, handling and extraction of useful knowledge from these data sources is currently one of the most popular and hot topics in computing research.

In this context, Big Data is a popular phenomenon which aims to provide an alternative to traditional solutions database and data analysis, leading to a revolution not only in terms of technology but also in business. It is not just about storage of and access to data, Big Data solutions aim to analyze data in order to make sense of that data and exploiting its value. One of the current main challenges in Data Mining related to Big Data problems is to find adequate approaches to analyze massive data online (or data streams) [43]. Due to classification methods requires from a previous labelling process, these methods need high efforts for real-time analysis. However, due to unsupervised techniques do not need this previous process, clustering becomes a promising field for real-time analysis. Clustering is perhaps one of the most popular approaches used in *unsupervised machine learning* and in Data Mining [25]. It is used to find hidden information or patterns in an unlabelled dataset and has several applications related to biomedicine, marketing [24], or image segmentation [50] amongst others. Clustering algorithms provide a large number of methods to search for "blind" patterns in data, some of these approaches are based on Bio-inspired methods such as evolutionary computation [21,27], swarm intelligence [9] or neural networks amongst others.

In the last years, and due to the fast growing of large Big Data-based problems, new challenges are appearing in previous research areas to manage the new features and problems that these types of problems produce. New kinds of algorithms, as *online clustering* or *streaming clustering* are appearing to deal with the main problems related to Big Data domains. When data streams are analyzed, it is important to consider the analysis goal, in order to determine the best type of algorithm to be used. We could divide data stream analysis in two main categories:

– *Offline analysis*: we consider a portion of data (usually large data) and apply an offline clustering algorithm to analyze this data.
– *Online analysis*: the data are analyzed in real-time. These kinds of algorithms are constantly receiving new data instances and are not usually able to keep past information. The most relevant limitations of these systems are: the data order matters and can not be modified; the data can not be stored or re-analyzed during the process; the results of the analysis depend on the time

---

[2] http://aci.info/2014/07/12/the-data-explosion-in-2014-minute-by-minute-infographic/.

the algorithm has been stopped. The main problem of these algorithms is that they need a specific space to update the information. This reduces the possibilities of the new algorithm.

From our previous experience in different complex and industrial problems in different areas from Social Networks Analysis [6,7], Project Scheduling, Videogames, Music classification, Unmmaned Systems, or Bio-informatics, we have designed and developed several bioinspired algorithms for clustering or graph-based computing with the aim to handle Big Data-based problems. We can distinguish from two main types of algorithms, those that have combined evolutionary strategies (mainly genetic algorithms) [42,44] and the second ones which have been designed using swarm intelligence approaches (ant colonies optimization algorithms) [23].

5.2   Social-Based Analysis and Mining

With the large number and fast growing of Social Media systems and applications, Social-based applications for Data Mining, Data Analysis, Big Data computation, Social Mining, etc. has become an important and hot topic for a wide number of research areas [5]. Although there exists a large number of existing systems (e.g., frameworks, libraries or software applications) which have been developed, and currently are used in various domains and applications based on Social Media. The applications and their main technologies used are mainly based on Big Data, Cloud or Grid Computing. The concept of Ephemeral computing has been rarely considered.

Most of the current challenges under study in Social-based analysis and mining are related to the problem of efficient knowledge representation, management and discovery. Areas as Social Network Analysis (SNA), Social Media Analytics (SMA) and Big Data, have as main aims to track, trends discovery or forecasting, so methods and techniques from: Opinion Mining, Sentiment Analysis, Multimedia management or Social Mining are commonly used. For example, when anyone tries to analyze how a Social Network is evolving using a straightforward representation based on a graph, but ignoring the information flow between nodes the information extracted from this analysis will be very limited. Other simple example, based on SNA, is an application that could try to extract behavioral patterns among users connected to a particular social network without taking into account their connections, their strengthens, or how their relationships are evolving through time. Social Big Data analysis, instead, aims to study large-scale Web phenomena such as Social Networks from a holistic point of view, i.e., by concurrently taking into account all the socio-technical aspects involved in their dynamic evolution.

Previous domains could be joined into a more general application area named *Social Big Data*. This area, or application domain, comes from the joining efforts of two domains: Social Media and Big Data. Therefore, Social Big Data will be based on the analysis of very-large to huge amount of data, which could belong to several distributed sources, but with a strong focus on Social media. Hence,

Social Big Data analysis [12, 38] is inherently interdisciplinary and spans areas such as Data Mining, Machine Learning, Statistics, Graph Mining, Information Retrieval, Linguistics, Natural Language Processing, Semantic Web, Ontologies, or Big Data Computing, amongst others. Their applications can be extended to a wide number of domains such as health and political trending and forecasting, hobbies, e-business, cyber-crime, counter terrorism, time-evolving opinion mining, social network analysis, or human-machine interaction.

Taking into account the nature of Social Big Data sources and the necessary processes and methods that will be required for data processing, the knowledge models, and possibly the analysis and visualization techniques to allow discover meaningful patterns [29], the potential application of Eph-C features could generate a new kind of algorithms that would be suitably applied in ephemeral environments.

5.3   Artificial Intelligence in Computer Games and Ephemerality

The application of artificial/computational intelligence to games (game AI/CI) has seen major advancements in the last decade and has settled as a separate research field [32, 60]. One of the main focus of the research is to provide computers with the capacity to perform tasks that are believed to require human intelligence, and that results in a number of interesting sub-fields such as AI-assisted game design, computational narrative, procedural content generation, non-player-character (NPC) behavior learning, NPC affective computing, believable bots, social simulation, and player modeling, among others [36]. Many of the problems that arise in these areas require creativity [34] and cannot be solved just proficiently but also in a human-like style; many interactions and relations emerge naturally in games what creates a complex system that is usually not easy to understand by a human but that can provide interesting results from a human perspective [56]. Moreover, many games have an ephemeral nature, hard to manage computationally. Some game assets (i.e., game contents, NPC behavior/game AI, game goals and even game rules) can be seen as volatile in the sense that one cannot guarantee they occur again. Thus, it does make sense to consider creating them ephemerally.

Furthermore, the recent boom of casual games played in mobile devices provokes that both the design and gameplay of games demand resources that appear and evaporate continuously during the execution of a game. This precisely occurs in the so-called pervasive games (i.e., "games that have one or more salient features that expand the contractual magic circle of play spatially, temporally, or socially" [45]) where the gaming experience is extended out in the real world. Playing games in the physical world requires computations that should be executed on the fly in the user's mobile device and having into account that players can decide to join or drop out the game in each instant. This same situation happens in most of the multiplayer games.

But we should not focus our attention just to this specific genre of games as many areas of application for Eph-C can be easily found in the game universe. So, it is not unreasonable to think about the concept of ephemeral games as those games that can be only placed once or that expires in some way; one can find many reasons for their creation as for instance: economic intentions (e.g., the player will be supposed to demand extensions of the game in the future) or creative aspects (e.g., provide unique game experiences by playing a game with irreversible actions). In addition, one can think about ephemeral goals/events that have temporary existence in games as these appear (and disappear) as consequences of the actions and preferences of the players. These goals/events are usually secondary (as the main goal is well-defined and related with the primary story of the game) but help significantly to improve the game experience and thus they are critical to increase user satisfaction (incidentally, the maximum objective of games). Other issue to consider is the reversibility of player's actions: most games provide the option to save the current state to reload it later, basically implying that players do not face the consequences of their acts as they can go back to a previous state; while this is interesting (and desirable) in a number of games, it is also truth that it is inconvenient in certain types of games as in multiplayer on-line games (e.g., first-person-shooter, real-time strategy, or role-playing games, among others) where the actions of a player influence the universe of the game and thus affect other players; goals, players? alliances, and even rewards have to be rearranged according to the game progress what grants temporality to the nature of game. This transitory essence of games produces important problems that are difficult to manage computationally, and where and how to create the volatile features of a game is a question that remains opens and that Eph-C can help to solve/mitigate.

Other issue to consider is the energy consumption, specially from the hardware resources. Videogames are played on a wide set of platforms including game consoles, personal computers and mobile devices, and all of them consume a large amount of kilowatt-hours per gaming session (not to mention the energy use of the monitors or other pheripherical devices); moreover, even if the user leaves the game for a while (perhaps hours), the gaming plaftform can consume as much energy as in an active play. In mobile devices this issue particularly affects battery consumption. Can Eph-C help to manage the energy cosumption in gaming sessions without negatively affecting the player's game experience, that is to say, without turning off the game platform or enabling the automatic power-down feature built into the device? This is the question that remains open and provides interesting lines of research. Moreover, in games that demand physical activity (such as dance games or pervasive games) it is important to manage the physical effort of the player; in this context Eph-C might help to do it by arranging the ephemeral goals wiht this goal in mind.

5.4 Computational Creativity

Computational Creativity has gained attention in the last few years [40]. The idea is not to exclude human artists from creative processes, substituting them by computer algorithms, but instead to extend human creativity by computer aided processes. Although many approaches to the concept are possible, several models arising from the Computational Intelligence (CI) area have been developed in the last decade. These models include the possibility of human interaction within the algorithms [57]. Art, design and content generations are areas of interest for both CI and audiovisual industry [58].

Interactive EAs (IEAs) as well as Human-Based Computational Intelligence [31] are interesting starting points in the area. With the advent of IEAs, and their possibilities for developing creativity (applications can be found in music composition [15], videogames plot induction and story generation [41], automatic poetry [39], etc.), a new problem arose: user fatigue due to multiple evaluations required when the fitness function is substituted by a human in charge of aesthetic evaluations. It is thus required to improve available IEAs so that new autonomous software tools can be developed and this implies a better understanding of human creative processes. Recently, a new proposal based on EAs has tried to analyze creativity when developed by human artists [17]. Results have provided clues that may lead in the future to new genetic operators or algorithms. Yet, high computational costs are associated with computer based creative processes [52], and distributed infrastructures are required. Among the possibilities, Eph-C-based models share some features with the way a team of artists can collaborate when developing evolutionary art following evolutionary approaches: highly asynchronous processes; completely distributed and frequently isolated way of working with some interaction along the work: artists work alone in their particular ?atelier?, and sometimes share their ideas in collective ephemeral activities, such as public exhibition, where the interaction with the audience and critics is exhibited. Therefore, ephemeral behavior is inherent to the way artists work and react to colleagues and the public. Previous models [17] could be thus studied from this point of view to improve existing methodologies.

6   Conclusions

Ephemeral computation provides an interesting new, and promising, research area with significant differences when it is compared against other areas as grid computing, or traditional distributed computing. Although Eph-C presents some features close to volunteer computing or amorphous computing, the combination of their main features: *inclusion, asynchrony, resilience, emergence, and self-adaptation*, defines it more precisely.

Therefore, the main focus of Eph-C is different from the above approaches. Rather than trying to build layers onto the network of ephemeral resources in order to "hide" their transient nature and provide the illusion of a virtual stable environment, Eph-C applications are fully aware of the nature of the computational landscape and are specifically built to live (and optimize their performance) in this realm.

Related to the application of traditional methods and techniques from Machine Learning to Big Data problems, our previous experience has shown the high performance that bioinspired algorithms can achieve in huge, open and dynamic problems, showing how bioinspired approaches can be used to improve the performance of unsupervised approaches. In the near future, and taking into account the new restrictions and features imposed by Eph-C environments, a new suit of algorithms able to efficiently handle the new challenges in data management and knowledge discovery in large Big Data-based problems will be studied and analyzed.

Acknowledgements. This work is supported by MINECO project EphemeCH (TIN2014-56494-C4-1-P, -2-P, -3-P and -4-P) – Check http://blog.epheme.ch.

References

1. Abelson, H., Allen, D., Coore, D., Hanson, C., Homsy, G., Knight Jr., T.F., Nagpal, R., Rauch, E., Sussman, G.J., Weiss, R.: Amorphous computing. Commun. ACM **43**(5), 74–82 (2000)
2. Álvarez, J.D., Colmenar, J.M., Risco-Martín, J.L., Lanchares, J., Garnica, O.: Optimizing l1 cache for embedded systems through grammaticalevolution. Soft Comput. **20**, 1–15 (2015)
3. Armbrust, M., Fox, A., Griffith, R., Joseph, A.D., Katz, R., Konwinski, A., Lee, G., Patterson, D., Rabkin, A., Stoica, I., et al.: A view of cloud computing. Commun. ACM **53**(4), 50–58 (2010)
4. Babaoglu, O., Jelasity, M., Montresor, A., Fetzer, C., Leonardi, S., Moorsel, A., Steen, M.: The self-star vision. In: Babaoglu, O., Jelasity, M., Montresor, A., Fetzer, C., Leonardi, S., Moorsel, A., Steen, M. (eds.) SELF-STAR 2004. LNCS, vol. 3460, pp. 1–20. Springer, Heidelberg (2005). doi:10.1007/11428589_1
5. Bello-Orgaz, G., Jung, J.J., Camacho, D.: Social big data: recent achievements and new challenges. Inf. Fusion **28**, 45–59 (2016)
6. Bello-Orgaz, G., Menéndez, H., Okazaki, S., Camacho, D.: Combining social-based data mining techniques to extract collective trends from twitter. Malaysian J. Comput. Sci. **27**(2), 95–111 (2014)
7. Bello-Orgaz, G., Menendez, H.D., Camacho, D.: Adaptive k-means algorithm for overlapped graph clustering. Int. J. Neu. Syst. **22**(05), 1250018 (2012)
8. Berl, A., Gelenbe, E., Di Girolamo, M., Giuliani, G., De Meer, H., Dang, M.Q., Pentikousis, K.: Energy-efficient cloud computing. Comput. J. **53**(7), 1045–1051 (2010)
9. Bonabeau, E., Dorigo, M., Theraulaz, G.: Swarm Intelligence: From Natural to Artificial Systems. Oxford University Press Inc., New York (1999)
10. Borella, M.S., Grabelsky, D., Nessett, D.M., Sidhu, I.S.: Method and system for locating network services with distributednetwork address translation. US Patent 6,055,236 (2000)
11. Bunse, C., Hopfner, H., Mansour, E., Roychoudhury, S.: Exploring the energy consumption of data sorting algorithms inembedded and mobile environments. In: Tenth International Conference on Mobile Data Management: Systems, Services and Middleware, MDM 2009, pp. 600–607. IEEE (2009)
12. Cambria, E., Rajagopal, D., Olsher, D., Das, D.: Big social data analysis. Big Data Comput. **13**, 401–414 (2013)
13. Cantú-Paz, E.: A survey of parallel genetic algorithms. Calculateurs paralleles reseaux et systems repartis **10**(2), 141–171 (1998)
14. Cotta, C., Sevaux, M., Sörensen, K. (eds.): Adaptive and Multilevel Metaheuristics. SCI, vol. 136. Springer, Heidelberg (2008)

15. Diaz-Jerez, G.: Composing with melomics: delving into the computational world formusical inspiration. Leonardo Music J. **21**, 13–14 (2011)
16. Eiben, A.E.: Evolutionary computing and autonomic computing: shared problems, shared solutions? In: Babaoglu, O., Jelasity, M., Montresor, A., Fetzer, C., Leonardi, S., Moorsel, A., Steen, M. (eds.) SELF-STAR 2004. LNCS, vol. 3460, pp. 36–48. Springer, Heidelberg (2005). doi:10.1007/11428589_3
17. Fernández de Vega, F., Navarro, L., Cruz, C., Chavez, F., Espada, L., Hernandez, P., Gallego, T.: Unplugging evolutionary algorithms: on the sources of novelty and creativity. In: IEEE Congress on Evolutionary Computation (CEC), pp. 2856–2863. IEEE (2013)
18. Flinn, J., Satyanarayanan, M.: Powerscope: a tool for profiling the energy usage of mobile applications. In: Second IEEE Workshop on Mobile Computing Systems and Applications, Proceedings, WMCSA 1999, pp. 2–10. IEEE (1999)
19. Fong, K.F., Hanby, V.I., Chow, T.-T.: HVAC system optimization for energy management by evolutionary programming. Energy Buildings **38**(3), 220–231 (2006)
20. Frei, R., McWilliam, R., Derrick, B., Purvis, A., Tiwari, A., DI Marzo Serugendo, G.: Self-healing and self-repairing technologies. Int. J. Adv. Manuf. Technol. **69**(5–8), 1033–1061 (2013)
21. Goldberg, D.E.: Genetic Algorithms in Search, Optimization and Machine Learning, 1st edn. Addison-Wesley Longman Publishing Co., Inc., Reading (1989)
22. Lombraña González, D., Jiménez Laredo, J.L., Fernández de Vega, F., Merelo Guervós, J.J.: Characterizing fault-tolerance of genetic algorithms in desktop grid systems. In: Cowling, P., Merz, P. (eds.) EvoCOP 2010. LNCS, vol. 6022, pp. 131–142. Springer, Heidelberg (2010). doi:10.1007/978-3-642-12139-5_12
23. Gonzalez-Pardo, A., Camacho, D.: Solving project scheduling problems through swarm-based approaches. Int. J. BioInspired Comput. (IJBIC) (2015, inpress)
24. Haider, P., Chiarandini, L., Brefeld, U.: Discriminative clustering for market segmentation. In: Proceedings of the 18th ACM SIGKDD international conferenceon Knowledge discovery and data mining, KDD 2012, pp. 417–425. ACM, New York (2012)
25. Han, J., Kamber, M.: Data Mining: Concepts and Techniques. Morgan Kaufmann, San Francisco (2006)
26. Harmon, R.R., Auseklis, N.: Sustainable it services: assessing the impact of green computing practices. In: Portland International Conference on Management of Engineering & Technology, PICMET 2009, pp. 1707–1717. IEEE (2009)
27. Holland, J.H.: Adaptation in Natural and Artificial Systems: An Introductory Analysis with Applications to Biology, Control and Artificial Intelligence. MIT Press, Cambridge (1992)
28. Huhns, M.N., Singh, M.P.: Service-oriented computing: key concepts and principles. IEEE Internet Comput. **9**(1), 75–81 (2005)
29. Kaisler, S., Armour, F., Espinosa, J.A., Money, W.: Big data: issues and challenges moving forward. In: 46th Hawaii InternationalConference on System Sciences (HICSS), pp. 995–1004. IEEE (2013)
30. Kamil, S., Shalf, J., Oliker, L., Skinner, D.: Understanding ultra-scale application communication requirements. In: Proceedings of the IEEE International Workload Characterization Symposium, 2005, pp. 178–187. IEEE (2005)
31. Kosorukoff, A.: Human based genetic algorithm. In: IEEE International Conference on Systems, Man, and Cybernetics, vol. 5, pp. 3464–3469. IEEE (2001)
32. Lara-Cabrera, R., Cotta, C., Fernández-Leiva, A.J.: A review of computational intelligence in rts games. In: IEEE Symposium on Foundations of Computational Intelligence, pp. 114-121. IEEE Press, Singapore (2013)

33. Laredo, J.L.J., Castillo, P.A., Mora, A.M., Merelo, J.J., Fernandes, C.: Resilience to churn of a peer-to-peer evolutionary algorithm. Int. J. High Performance Syst. Architect. **1**(4), 260–268 (2008)

34. Liapis, A., Yannakakis, G.N., Togelius, J.: Computational game creativity. In: Proceedings of the Fifth International Conference on Computational Creativity (ICCC 2014) (2014)

35. Lohr, S.: The age of big data. New York Times, 11 February 2012. Online. Accessed 5 Sept. 2014

36. Lucas, S.M., Mateas, M., Preuss, M., Spronck, P., Togelius, J., (eds.) Artificial and Computational Intelligence in Games, vol. 6. Dagstuhl Follow-Ups. Schloss Dagstuhl - Leibniz-Zentrum fuer Informatik (2013)

37. Lyytinen, K., Yoo, Y.: Ubiquitous computing. Commun. ACM **45**(12), 63–96 (2002)

38. Manovich, L.: Trending: the promises and the challenges of big social data. In: Debates in the Digital Humanities, pp. 460–475 (2011)

39. Manurung, H.: An evolutionary algorithm approach to poetry generation. PhD thesis, University of Edinburgh. College of Science and Engineering. School of Informatics (2004)

40. McCormack, J., D'Iverno, M.: Computers and Creativity. Springer, Heidelberg (2012)

41. McIntyre, N., Lapata, M.: Plot induction and evolutionary search for story generation. In: Proceedings of the 48th Annual Meeting of the Association for Computational Linguistics, pp. 1562–1572. Association for Computational Linguistics (2010)

42. Menéndez, H.D., Barrero, D.F., Camacho, D.: A genetic graph-based approach for partitional clustering. Int. J. Neural Syst. **24**(03) (2014a)

43. Menéndez, H.D., Otero, F.B., Camacho, D.: Extending the SACOC algorithm through the Nystrom method for bigdata analysis. Int. J. Bio-Inspired Comput. (2016, in press)

44. Menéndez, H.D., Otero, F.E.B., Camacho, D.: MACOC: a medoid-based ACO clustering algorithm. In: Dorigo, M., Birattari, M., Garnier, S., Hamann, H., Montes de Oca, M., Solnon, C., Stützle, T. (eds.) ANTS 2014. LNCS, vol. 8667, pp. 122–133. Springer, Heidelberg (2014). doi:10.1007/978-3-319-09952-1_11

45. Montola, M., Stenros, J., Waern, A.: Pervasive Games. Morgan Kaufmann, Boston (2009)

46. Network for Sustainable Ultrascale Computing. The future of ultrascale computing under study (2014). Online, Accessed 8 Sept. 2014

47. Nogueras, R., Cotta, C.: Studying fault-tolerance in island-based evolutionary and multimemetic algorithms. J. Grid Comput. (2015a). doi:10.1007/s10723-014-9315-6

48. Nogueras, R., Cotta, C.: Studying self-balancing strategies in island-based multimemetic algorithms. J. Comput. Applied Math. (2015b). doi:10.1016/j.cam.2015.03.047

49. Nogueras, R., Cotta, C.: Towards resilient multimemetic systems on unstable networks with complex topology. In: Papa, G. (ed.) Advances in Evolutionary Algorithm Research. Nova Science Pub. (2015c, in press)

50. Pascual, A., Barcéna, M., Merelo, J.J., Carazo, J.-M.: Application of the fuzzy Kohonen clustering network to biological macromolecules images classification. In: Mira, J., Sánchez-Andrés, J.V. (eds.) IWANN 1999. LNCS, vol. 1607, pp. 331–340. Springer, Heidelberg (1999). doi:10.1007/BFb0100500

51. Prasithsangaree, P., Krishnamurthy, P.: Analysis of energy consumption of RC4 and AES algorithms in wireless LANs. In: Global Telecommunications Conference, GLOBECOM 2003, vol. 3, pp. 1445–1449. IEEE (2003)
52. Reis, G., de Vega, F.F., Ferreira, A.: Automatic transcription of polyphonic piano music using genetic algorithms, adaptive spectral envelope modeling, and dynamic noise level estimation. IEEE Trans. Audio Speech Lang. Process. 20(8), 2313–2328 (2012)
53. Sarmenta, L.F., Hirano, S.: Bayanihan: building and studying web-based volunteer computing systems using Java. Future Gener. Comput. Syst. 15(5), 675–686 (1999)
54. Sharmin, M., Ahmed, S., Ahamed, S.I.: SAFE-RD (secure, adaptive, fault tolerant, and efficient resource discovery) in pervasive computing environments. In: International Conference on Information Technology: Coding and Computing, ITCC 2005, vol. 2, pp. 271–276. IEEE (2005)
55. Stutzbach, D., Rejaie, R.: Understanding churn in peer-to-peer networks. In: Proceedings of the 6th ACM SIGCOMM Conference on Internet Measurement, IMC 2006, pp. 189–202. ACM, New York (2006)
56. Sweetser, P.: Emergence in Games. Game Development. Charles River Media, Boston (2008)
57. Takagi, H.: Humanized computational intelligence with interactive evolutionary computation. In: Fogel, D.B., Robinson, C.J. (eds.) Computational Intelligence: The Experts Speak, pp. 207–218. Wiley (2003)
58. Togelius, J., Yannakakis, G.N., Stanley, K.O., Browne, C.: Search-based procedural content generation: a taxonomy and survey. IEEE Trans. Comput. Intell. AI Games 3(3), 172–186 (2011)
59. Wang, B., Bodily, J., Gupta, S.K.: Supporting persistent social groups in ubiquitous computing environments using context-aware ephemeral group service. In: Proceedings of the Second IEEE Annual Conference on Pervasive Computing and Communications, PerCom 2004, pp. 287–296. IEEE (2004)
60. Yannakakis, G., Togelius, J.: A panorama of artificial and computational intelligence in games. IEEE Trans. Comput. Intell. AI Games 7(4), 317–335 (2015)

Author Index

Printed in the United States
By Bookmasters